Studies in Computational Intelligence

Volume 893

Series Editor

Janusz Kacprzyk, Polish Academy of Sciences, Warsaw, Poland

The series "Studies in Computational Intelligence" (SCI) publishes new developments and advances in the various areas of computational intelligence—quickly and with a high quality. The intent is to cover the theory, applications, and design methods of computational intelligence, as embedded in the fields of engineering, computer science, physics and life sciences, as well as the methodologies behind them. The series contains monographs, lecture notes and edited volumes in computational intelligence spanning the areas of neural networks, connectionist systems, genetic algorithms, evolutionary computation, artificial intelligence, cellular automata, self-organizing systems, soft computing, fuzzy systems, and hybrid intelligent systems. Of particular value to both the contributors and the readership are the short publication timeframe and the world-wide distribution, which enable both wide and rapid dissemination of research output.

Indexed by SCOPUS, DBLP, WTI Frankfurt eG, zbMATH, SCImago.

All books published in the series are submitted for consideration in Web of Science.

More information about this series at http://www.springer.com/series/7092

Christophe Sabourin · Juan Julián Merelo ·
Alejandro Linares Barranco · Kurosh Madani ·
Kevin Warwick

Editors

Computational Intelligence

International Joint Conference, IJCCI 2018
Seville, Spain, September 18–20, 2018
Revised Selected Papers

 Springer

Editors
Christophe Sabourin
IUT Sénart, University Paris-Est Cretei
Lieusaint, France

Juan Julián Merelo
Computer Architecture and Technology
University of Granada
Granada, Spain

Alejandro Linares Barranco
ETSI Informática
Sevilla, Spain

Kurosh Madani
University of Paris-EST Créteil (UPEC)
Créteil, France

Kevin Warwick
University of Reading and Coventry
University
Coventry, UK

ISSN 1860-949X ISSN 1860-9503 (electronic)
Studies in Computational Intelligence
ISBN 978-3-030-64733-9 ISBN 978-3-030-64731-5 (eBook)
https://doi.org/10.1007/978-3-030-64731-5

This Springer imprint is published by the registered company Springer Nature Switzerland AG
The registered company address is: Gewerbestrasse 11, 6330 Cham, Switzerland

Organization

Conference Chair

Kurosh Madani, University of Paris-EST Créteil (UPEC), France

Program Co-chairs

Christophe Sabourin, IUT Sénart, University Paris-Est Creteil (UPEC), France
Juan Julián Merelo, University of Granada, Spain
Alejandro Linares Barranco, ETSI Informática, Spain

Program Committee

Jean-Jacques Mariage, Laboratoire d'Informatique Avancée de Saint-Denis and Université Paris 8,—Select a Country
Salmiah Ahmad, International Islamic University Malaysia, Malaysia
Otar Akanyeti, Aberystwyth University, UK
Jesús Alcalá-Fdez, University of Granada, Spain
Richard Allmendinger, University of Manchester, UK
Majid Amirfakhrian, Central Tehran Branch, Islamic Azad University, Iran, Islamic Republic of
Davide Anguita, University of Genoa, Italy
Michela Antonelli, University of Pisa, Italy
Vijayan Asari, University of Dayton, USA
Sansanee Auephanwiriyakul, Chiang Mai University, Thailand
Dalila B.M.M. Fontes, Faculdade de Economia and LIAAD-INESC TEC, Universidade do Porto, Portugal
Thomas Baeck, Leiden University, Netherlands

Stefka Fidanova, Bulgarian Academy of Sciences, Bulgaria
Leonardo Franco, Universidad de Málaga, Spain
Yoshikazu Fukuyama, Meiji University, Japan
Jonathan Garibaldi, University of Nottingham, UK
Nizami Gasilov, Baskent University, Turkey
Alexandros Giagkos, Aston University, UK
David Gil Mendez, University of Alicante, Spain
Vladimir Golovko, Brest State Technical University, Belarus
Michèle Gouiffès, Institut D'Electronique Fondamentale (IEF) CNRS 8622 University Paris-Sud 11, France
Sarah Greenfield, De Montfort University, UK
Hazlina Hamdan, Universiti Putra Malaysia, Malaysia
Lutz Hamel, University of Rhode Island, USA
Oussama Hamid, University of Nottingham, UK
Thomas Hanne, University of Applied Arts and Sciences Northwestern Switzerland, Switzerland
Rainer Heinrich Palm, AASS, Department of Technology, Örebro University, SE-70182 Örebro and Sweden,—Select a Country
Arturo Hernández—Aguirre, Centre for Research in Mathematics, Mexico
Chris Hinde, Loughborough University, UK
Wladyslaw Homenda, Warsaw University of Technology, Poland
Katsuhiro Honda, Osaka Prefecture University, Japan
Wei-Chiang Hong, Jiangsu Normal University, China
Alexander Hošovský, Technical University of Kosice, Slovak Republic
Gareth Howells, University of Kent, UK
Jiansheng Huang, Western Sydney University, Australia
Nor Husin, Universiti Putra Malaysia, Malaysia
Daniela Iacoviello, Sapienza Università di Roma, Italy
Yuji Iwahori, Chubu University, Japan
Dmitry Kangin, University of Exeter, UK
Iwona Karcz-Duleba, Wroclaw University of Science Technology, Poland
Uzay Kaymak, Eindhoven University of Technology, Netherlands
Christel Kemke, University of Manitoba, Canada
Etienne Kerre, Ghent University, Belgium
Georges Khazen, Lebanese American University, Lebanon
Ahmed Kheiri, Lancaster University, UK
DaeEun Kim, Yonsei University, Korea, Republic of
Frank Klawonn, Ostfalia University of Applied Sciences, Germany
László Kóczy, Budapest University of Technology and Economics, Hungary
Mario Köppen, Kyushu Institute of Technology, Japan
Vladik Kreinovich, University of Texas at El Paso, USA
Ondrej Krejcar, University of Hradec Kralove, Czech Republic
Pavel Krömer, VSB Ostrava, Czech Republic
Yau-Hwang Kuo, National Cheng Kung University, Taiwan, Republic of China
Dario Landa-Silva, University of Nottingham, UK

Gary Parker, Connecticut College, USA
David A. Pelta, University of Granada, Spain
Parag Pendharkar, Pennsylvania State University, USA
Irina Perfilieva, University of Ostrava, Czech Republic
Valentina Plekhanova, University of Sunderland, UK
Radu-Emil Precup, Politehnica University of Timisoara, Romania
Amaryllis Raouzaiou, National Technical University of Athens, Greece
Joaquim Reis, ISCTE, Portugal
Mateen Rizki, Wright State University, USA
Antonello Rizzi, Università di Roma "La Sapienza", Italy
Olympia Roeva, Institute of Biophysics and Biomedical Engineering, Bulgarian
Academy of Sciences, Bulgaria
Roseli Romero, University of São Paulo, Brazil
Neil Rowe, Naval Postgraduate School, USA
Suman Roychoudhury, Tata Consultancy Services, India
Christophe Sabourin, IUT Sénart, University Paris-Est Creteil (UPEC), France
Miguel Sanz-Bobi, Comillas Pontifical University, Spain
Robert Schaefer, AGH University of Science and Technology, Poland
Alon Schclar, Academic College of Tel-Aviv Yaffo, Israel
Tjeerd Scheper, Oxford Brookes University, UK
Christoph Schommer, University Luxembourg, Campus Belval, Maison du Nombre,
Luxembourg
Daniel Schwartz, Florida State University, USA
Shahnaz Shahbazova, Azerbaijan Technical University, Azerbaijan
Yabin Shao, Chongqing University of Posts and Telecommunications, China
Qiang Shen, Aberystwyth University, UK
Patrick Siarry, University Paris 12 (LiSSi), France
Andrzej Skowron, Systems Research Institute, Polish Academy of Sciences, Poland
Catherine Stringfellow, Midwestern State University, USA
Mu-Chun Su, National Central University, Taiwan, Republic of China
Johan Suykens, KU Leuven, Belgium
Norikazu Takahashi, Okayama University, Japan
Tatiana Tambouratzis, University of Piraeus, Greece
Yi Tang, Yunnan University of Nationalities, China
C. Tao, National I-Lan University, Taiwan, Republic of China
Guy Theraulaz, Centre National de la Recherche Scientifique, France
Philippe Thomas, Université de Lorraine, France
Vicenc Torra, University of Skövde, Sweden
Juan-Manuel Torres-Moreno, Ecole Polytechnique de Montréal, Canada
Dat Tran, University of Canberra, Australia
Carlos Travieso-González, Universidad de Las Palmas de Gran Canaria, Spain
Krzysztof Trojanowski, Uniwersytet Kardynała Stefana Wyszyńskiego, Poland
Tan Tse Guan, Universiti Malaysia Kelantan, Malaysia
Elio Tuci, Middlesex University, UK

Alexander Tulupyev, St. Petersburg Institute for Informatics and Automation of the Russian Academy of Sciences (SPIIRAS), Russian Federation
Jessica Turner, Georgia State University, USA
Andrei Utkin, INOV—INESC Inovação, Portugal
Lucia Vacariu, Technical University of Cluj Napoca, Romania
Arjen van Ooyen, VU University Amsterdam, Netherlands
Wen-June Wang, National Central University, Taiwan, Republic of China
Guanghui Wen, Southeast University, China
Li-Pei Wong, Universiti Sains Malaysia, Malaysia
Jian Wu, School of Economics and Management, Shanghai Maritime University, China
Yiyu Yao, University of Regina, Canada
Chung-Hsing Yeh, Monash University, Australia
Hao Ying, Wayne State University, USA
Umi Yusof, Universiti Sains Malaysia, Malaysia
Slawomir Zadrozny, Polish Academy of Sciences, Poland
Cleber Zanchettin, Federal University of Pernambuco, Brazil
Hans-Jürgen Zimmermann, ELITE (European Laboratory for Intelligent Techniques Engineering), Germany.

Invited Speakers

Keeley Crockett, Manchester Metropolitan University, UK
Thomas Villmann, University of Applied Sciences Mittweida, Germany
Oscar Cordón, University of Granada, Spain
Humberto Bustince, Public University of Navarra, Spain

Preface

The present book includes extended and revised versions of a set of selected papers from the 10th International Joint Conference on Computational Intelligence (IJCCI 2018), held in Seville, Spain, from 18 to 20 September 2018.

IJCCI 2018 received 47 paper submissions from 23 countries, of which 15% were included in this book. The papers were selected by the event chairs based on a number of criteria that include the classifications and comments provided by the program committee members, the session chairs' assessment and also the program chairs' global view of all papers included in the technical program. The authors of selected papers were then invited to submit a revised and extended version of their papers having at least 30% innovative material and results.

The purpose of the International Joint Conference on Computational Intelligence—IJCCI—is to bring together researchers, engineers and practitioners interested in the field of Computational Intelligence both from theoretical and application perspectives. Four simultaneous tracks have been held covering foremost aspects of Computational Intelligence, namely, "evolutionary computation", "fuzzy computation", "neural computation" and "cognitive and hybrid systems". The connection of these areas in all their wide range of approaches and applications forms the International Joint Conference on Computational Intelligence.

If the diversity and complementarily of selected papers of this book contribute to the understanding of relevant trends of current research on Computational Intelligence, they also draw attention to recent pertinent achievements focusing a number of leading and concrete facets of this appealing field such as the field of Artificial Intelligence at large, Fuzzy Information Processing and Fusion in Text Mining, Speech and Signal Processing, application of Reinforcement Learning and Deep Learning algorithms in pattern and Gesture Recognition and finally, the practice of Artificial Neural Networks and Decision Trees in Psychological Profiling or Anomaly Detection.

We would like to thank all the authors for their valuable contributions without forgetting the precious work accomplished by reviewers who have evaluated the selected papers and greatly helped ensuring the quality of this publication.

Lieusaint, France   Christophe Sabourin
Granada, Spain   Juan Julián Merelo
Sevilla, Spain   Alejandro Linares Barranco
Créteil, France   Kurosh Madani
Coventry, UK   Kevin Warwick
September 2018

Contents

A Method for Selecting Suitable Records Based on Fuzzy Conformance and Aggregation Functions

Miljan Vučetić and Miroslav Hudec

Abstract Searching for suitable entities in a datasets is still a challenging task, because the entities' attributes are often expressed by various data types including numerical, categorical, and fuzzy data. In addition, an attribute in a dataset may convey different data types for diverse records. In the query, the user may explain requirements by a different data type comparing with the one stored in a dataset, i.e. by linguistic term(s), whereas the respective attribute in a dataset is recorded as a real number and vice versa. Further, the user may provide complex preferences among atomic conditions. In this paper, we propose a robust framework capable to manage user requirements and match them with records in a dataset. The former is solved by the conformance measure, whereas for the latter different aggregation functions belonging to the conjunctive, averaging and hybrid classes have been suggested to cover particular aggregation needs like coalitions among atomic predicates and quantified conditions. The proposed method can be applied for selecting suitable records from any dataset containing numerical, categorical, fuzzy and binary data. The important characteristic of the presented method is the efficient applicability in the mentioned data collections, because conformance measure can manage different data types in the same manner. Finally, we discuss benefits, drawbacks and outline further activities.

Keywords Similarity · Conformance measure · Conjunction · Averaging functions · Hybrid functions · Quantified fuzzy aggregation

M. Vučetić (✉)
Vlatacom Institute of High Technologies, 5 Milutina Milankovića Blvd, Belgrade, Serbia
e-mail: miljan.vucetic@vlatacom.com

M. Hudec
Faculty of Economic Informatics, University of Economics, Dolnozemská cesta 1, Bratislava, Slovakia
e-mail: miroslav.hudec@euba.sk

© The Author(s), under exclusive license to Springer Nature Switzerland AG 2021
C. Sabourin et al. (eds.), *Computational Intelligence*, Studies in Computational Intelligence 893, https://doi.org/10.1007/978-3-030-64731-5_1

1

1 Introduction

In retrieving the suitable entities (customers, products, territorial units, etc.) from datasets, users usually consider a higher number of atomic conditions which desired entities should met. Users require that the search process provides them with sensible responses to their requests [1]. A query should also return records which partially meet the user's requirements, especially when no record ideally matches the requirement.

In a dataset, attributes' values can be stored by a variety of data types and may be heterogeneous, i.e. values of one attribute may be stored for some records as numeric, whereas for others as fuzzy or categorical data. Further, users can express their expectations linguistically, categorically or numerically. Thus, it is a challenge to envelope a mixture of data types (numerical, categorical, binary, and fuzzy data) in a single query. Moreover, users' preferences might be expressed by a different data type comparing with the one used in a dataset, i.e. a user may explain that the desired flat distance to the nearest grocery shop is very short or short, whereas this attribute is recorded as a real number greater than 0. In the opposite case, a user may say that the expected distance to the nearest train stop should be within 200 m, but in a dataset the distance is expressed linguistically by one term from the following set of labels: {very short walking distance, short walking distance, medium walking distance, long walking distance, beyond walking distance}.

These requirements make application of usual fuzzy query approaches such as: FQUERY [2], FQL [3], SQLf [4], GLC [5], FSQL [6], PFSQL [7] and their further extensions, hard. Hence, the promising alternative is augmenting them by conformance measures initially developed for calculating fuzzy functional dependencies [8, 9]. In this work, the definition of conformance is based on the fuzzy sets and proximity relation proposed in [10].

The overall query condition may consist of higher number of atomic ones (e.g. features of products which should be met). In the case of conjunction, it might cause the so-called *empty answer problem* [11], i.e. not a single record is selected. The simplest solution, excluding one or several atomic conditions is not the solution, because the user has its own reason for including the particular set of condition. The possible solutions are quantified queries, where majority of atomic conditions should be met [12], and relaxing atomic conditions [13]. Relaxing query condition is a complex task of recognizing the most suitable atomic predicates for the relaxation, and keeping the semantic meaning as close as possible to the initial query [13].

The former does not divide atomic conditions into the hard (must be imperatively met) and soft (it is nice if they are met as well), i.e. a record is the solution even if it does not meet one of the atomic conditions, regardless the importance of this condition. The possible solution is dividing atomic conditions into the hard and soft sets of constraints and quantifying them as is proposed in [14, 15]. In addition, users can express preferences among atomic conditions by other ways: equal preferences, weights, coalitions among atomic conditions, etc.

This study is the extended work of the preliminary results published at the IJCCI 2018 conference [16]. The first part examines calculating conformances initially developed in [17] and recently applied in recommending less-frequently purchased products [18]. In this extended version, we explain how conformance equation is applied to cover multiple scenarios in heterogeneous datasets including numerical, categorical, binary and fuzzy data. To demonstrate applicability we describe various ways to calculate how conformant are records in a dataset with the user's expectations. The second part of this study examines aggregations of atomic conditions in order to cover the expected preferences among attributes raised by users: equally important atomic conditions managed by conjunction and averaging functions, quantified conditions, merging hard and soft conditions and managing cases when several atomic conditions create so-called coalitions (i.e. the contribution of one atomic condition might be small, but significantly rise when other atomic conditions are significantly satisfied). The third part covers discussion focused on the examples, benefits, issues, possible future research directions and concluding messages.

2 Conformance Measures

When calculating similarity among attribute's values, the fuzzy conformance-based approach has been shown as a suitable choice. It is straightforwardly used for matching complex user requirements with records in a dataset when heterogeneous data types are considered [18].

The conformance measure is used to compare expected and existing values of particular attributes. In this sense, the value of conformance, which belongs to the [0, 1] interval is reasonable for measuring how the user's requirements and items in the dataset match. Therefore, amongst many methods, this approach provides universal way for comparing given crisp, categorical and fuzzy data appearing in user preferences and attributes' values. Although data may be heterogeneous, we are able to straightforwardly measure the similarity between user requirements and item features by conformance equation presented in [10] which is based on the fuzzy sets and proximity relation:

$$C\left(X_i[t_u, t_j]\right) = \min\left(\mu_{tu}(x_i)\mu_{tj}(x_i), s\left(t_u(X_i), t_j(X_i)\right)\right) \tag{1}$$

where C is a fuzzy conformance of attribute X_i defined on the domain D_i between user requirement t_u and record t_j in a dataset, s is a proximity relation and $\mu_{tu}(X_i)$ and $\mu_{tj}(X_i)$ are membership degrees of user preferred value and value in a dataset, respectively.

When we analyse fuzzy data, it is necessary to answer how fuzzy value B belongs to the fuzzy set A (e.g. *price about 1000* belongs to the fuzzy set *medium price*). This is realized by the possibility measure defined as [19, 20]:

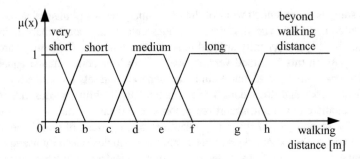

Fig. 1 An example of fuzzified attribute *walking distance* [16]

$$Poss(B, A) = \sup_{x \in X}[t(A(x), B(x))] \qquad (2)$$

where X is a universe of discourse and t is a t-norm. In practice, minimum t-norm is used. This equation is used to get membership degree when fuzzy data appears in user requirements and item features in a dataset.

Data transformation in the proposed method is required operation for selecting suitable records due to heterogeneous data types in queries and datasets. We propose a universal method based on the fuzzification of attributes domains. For instance, the attribute *walking distance* is fuzzified into several fuzzy sets, as shown in Fig. 1 [16]. The categorical and binary attributes are considered as fuzzy singletons in this data transformation step.

The equation for fuzzy conformance relies on proximity relations created for each attribute domain. These relations are reflexive and symmetric and do not meet the constraint of max–min transitivity as similarity relation does [21]. From the application point of view, the properties of proximity relation provide us with more flexibility to express similarity between linguistic terms.

In the data transformation step, it is necessary to define proximity relation on the scalar attribute domains. The same is applied on the fuzzified domains for numerical attributes. Specifically, by employing fuzzy sets for domain partitions, it is possible to describe similarities between mixed data types. Algorithms [18, 22, 23] calculate the intensity of compatibility between desired value and values of each record (item) in a dataset.

For example, the walking distance to the nearest grocery shop is in our case fuzzified as *very short walking distance, short walking distance, medium walking distance, long walking distance, beyond walking distance* as illustrated in Fig. 1. This approach enables flexibility in comparing numerical data and linguistic terms as is shown in example below. The same holds for the other attributes. For simplicity reasons, these linguistic terms are mathematically formalized by liner membership functions. Table 1 shows the proximity relation among aforementioned fuzzy sets.

Let us observe the following example of searching for a suitable flat. The user prefers walking distance (attribute A_1) $t_u(Walk\_Dist)$ of *less than 200 m*. Membership

Table 1 The proximity relation over the domain of the walking distance attribute, where wd stands for walking distance [16]

s_{wd}	Very short wd	Short wd	Medium wd	Long wd	Beyond wd
Very short wd	1	0.90	0.50	0.10	0
Short wd		1	0.80	0.25	0
Medium wd			1	0.85	0.45
Long wd				1	0.65
Beyond wd					1

degree to the fuzzy set *very short walking distance* is $\mu_{tu}(Walk\_Dist) = 1$, using Eq. (2). For each pair of user requirement and item in the dataset, we use Eq. (1) in order to calculate the intensity of similarity.

In the case of $t_1(Walk\_Dist) = 215$ m, the membership degree to the fuzzy set *very* short walking distance is $\mu_{t1}(Walk\_Dist) = 0.85$ and s(very short, very short) $= 1$, when parameters $a = 200$ and $b = 300$ in Fig. 1:

$$C(Walk\_Dist[t_u, t_1]) = \min(\mu_{tu}(Walk\_Dist), \mu_{t1}(Walk\_Dist),$$
$$s(t_u(Walk\_Dist), t_1(Walk\_Dist)))$$
$$= \min(1, 0.85, 1) = 0.85$$

The conformance of t_u and $t_2(Walk_Dist) = $ around 670 m (membership degree to the fuzzy set *medium walking distance* is $\mu_{t2}(Walk\_Dist) = 0.70$ using Eq. (2) where $c = 600$, $d = 700$, Fig. 1) is given as:

$$C(Walk\_Dist[t_u, t_2]) = \min(\mu_{tu}(Walk\_Dist), \mu_{t2}(Walk\_Dist),$$
$$s(t_u(Walk\_Dist), t_2(Walk\_Dist)))$$
$$= \min(1, 0.70, 0.50) = 0.50$$

The conformance of t_u and $t_3(Walk\_Dist)$, where t_3 contains linguistic term *long wd* is:

$$C(Walk\_Dist[t_u, t_3]) = \min(\mu_{tu}(Walk\_Dist), \mu_{t3}(Walk\_Dist),$$
$$s(t_u(Walk\_Dist), t_3(Walk\_Dist)))$$
$$= \min(1, 1, 0.10) = 0.10$$

It should be noted that conformance may be zero. For example, $C(Walk\_Dist[t_u, t_4])$ between t_u and $t_4(Walk\_Dist) = 2130$ m (membership degree to the fuzzy set *beyond walking distance* is $\mu_{t4}(Walk\_Dist) = 1$, when $h = 2100$ m in Fig. 1) for s(very short, beyond) $= 0$ from Table 1 is calculated as follows:

$$C(Walk\_Dist[t_u, t_4]) = \min(\mu_{tu}(Walk\_Dist),$$
$$\mu_{t3}(Walk\_Dist), s(t_u(Walk\_Dist), t_4(Walk\_Dist)))$$
$$= \min(1, 1, 0) = 0$$

Fig. 2 An example of
membership functions of a
singleton fuzzy sets

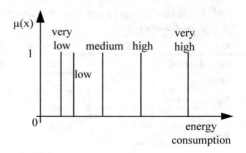

Table 2 The proximity relation over the domain of the energy consumption attribute

s_{en}	Very low	Low	Medium	High	Very high
Very low	1	0.95	0.70	0.40	0.10
Low		1	0.85	0.65	0.25
Medium			1	0.80	0.50
High				1	0.75
Very high					1

Obviously, the conformance of t_u and t_5, where t_5 contains numerical value of 195 m is 1. These conformances are shown in Table 4, for attribute A_1.

Similarly, we calculate conformances for the other attributes. For instance, attribute A_2 is energy consumption expressed by linguistic terms and therefore defined as categorical attribute. The linguistic terms are shown in Fig. 2. The proximity relation is defined in the same way as for the terms of the walking distance attribute to describe the similarity among the categorical values as is shown in Table 2.

Figure 2 shows how categorical attributes are fuzzifed in order to apply fuzzy conformance measures. The same principle is applied for binary (two-value) attributes.

The following example using categorical attribute A_2 shows calculation of fuzzy conformance:

$$C(Energy\_Cn[t_u, t_3]) = \min(\mu_{tu}(Energy\_Cn), \mu_{t3}(Energy\_Cn),$$
$$s(t_u(Energy\_Cn), t_3(Energy\_Cn)))$$
$$= \min(1, 1, 0.65) = 0.65$$

where $t_u(Energy\_Cn) = low$ and $t_3(Energy\_Cn) = medium$ are fuzzy singletons (Fig. 2) and $s(low, medium) = 0.65$.

The same holds in the following illustration for record t_5 of attribute A_2:

$$C(Energy\_Cn[t_u, t_5]) = \min(\mu_{tu}(Energy\_Cn), \mu_{t5}(Energy\_Cn),$$
$$s(t_u(Energy\_Cn), t_5(Energy\_Cn)))$$
$$= \min(1, 1, 0.25) = 0.25$$

where $t_u(Energy\_Cn) = low$ and $t_5(Energy\_Cn) = high$ and $s(low, high) = 0.25$.

The conformance on binary data usually assumes value 0, when the proximity between *Yes* and *No* is 0. For instance, attribute A_4 is *presence of the elevator* in the block of flats and therefore appears as a binary attribute in the dataset. When the user declares that there is no proximity between presence and non-presence, then binary values appear in Table 4.

Theoretically, the proximity can be greater than 0, when these two opposite cases are not fully exclusive for users. Let us observe the attribute *presence of balcony* in the following example:

$$C(Balcony[t_u, t_1]) = \min(\mu_{tu}(Balcony), \mu_{t1}(Balcony),$$
$$s(t_u(Balcony), t_1(Balcony)))$$
$$= \min(1, 1, 0.20) = 0.20$$

where $t_u(Balcony) = $ Yes and $t_1(Balcony) = $ No and $s(Yes, No) = 0.20$.

In this case flat without balcony is not fully excluded from the user's point of view. Due to proximity between two opposite values of 0.2, user might be still interested.

The next attribute may be an aggregated opinion about location on social networks from diverse respondents' categories (e.g., customers, journalists, domain experts). For such evaluation, the linguistic term sets (LTS) of different size might be used. An illustration of three linguistic term sets is as follows: $LTS_1 = $ {negative, neutral, positive}; $LTS_2 = $ {very negative, negative, neutral, positive, very positive}; $LTS_3 = $ {very negative, negative, more neutral than negative, neutral, more neutral than positive, positive, very positive}. In this way, a less granular scale avoids response burden of "ordinal evaluators", whereas a finer scale avoids speculations of domain experts that the collecting opinions is not sufficiently detailed. In order to store such opinions into a dataset, the solution is conversion into the basic linguistic term set (BLST) consisting for instance of five fuzzy sets. More about transformations into BLST is in [24]. An example of BLST used for the hypothetical evaluation of attribute in a dataset representing opinion on social networks about flats is in Table 3.

Let us analyse following examples. The user prefers positive opinion about flat (attribute A_7 in Table 4):

$$C(SN[t_u, t_1]) = \min(\mu_{tu}(SN), \mu_{t1}(SN), s(t_u(SN), t_1(SN)))$$
$$= \min(1, 0.85, 0.85) = 0.85$$

where $t_u(SN) = $ positive, $t_1(SN) = $ neutral due to $\max(0, 0.25, 0.85, 0, 0) = 0.85$ from Table 3 and $s(positive, neutral) = 0.85$.

Table 3 An example of stored data using BLST for the domain of the opinion attribute

Flat	$\mu_{(very\_positive)}$	$\mu_{(positive)}$	$\mu_{(neutral)}$	$\mu_{(negative)}$	$\mu_{(very\_negative)}$
t_1	0	0.25	0.85	0	0
t_2	0	0	0	0.6	0.4
t_3	0.158	0.842	0	0	0
t_4	0.752	0.248	0	0	0
t_5	0	0	0.596	0.404	0

Table 4 Fuzzy conformances of attributes A_1 to A_7 between user preferences expressed by vector of ideal values t_u and records t_1 to t_5 [16]

Rec	$C(A_1[t_u, t_j])$	$C(A_2[t_u, t_j])$	$C(A_3[t_u, t_j])$	$C(A_4[t_u, t_j])$	$C(A_5[t_u, t_j])$	$C(A_6[t_u, t_j])$	$C(A_7[t_u, t_j])$
t_1	0.85	0.85	0.85	1.00	0.85	0.85	0.85
t_2	0.50	0.25	0.26	1.00	0.29	0.24	0.27
t_3	0.10	0.65	0.46	1.00	0.41	0.88	0.44
t_4	0.00	0.95	0.88	1.00	1.00	0.90	0.85
t_5	1.00	0.25	0.65	0.00	0.25	0.00	0.35

$$C(SN[t_u, t_2]) = \min(\mu_{tu}(SN), \mu_{t2}(SN), s(t_u(SN), t_2(SN)))$$
$$= \min(1, 0.60, 0.27) = 0.27$$

when $t_2(SN) =$ negative $(\max(0, 0, 0, 0.60, 0.40) = 0.60)$ and s(positive, negative) $= 0.27$.

This notion of fuzzy conformance is related to the calculated degree of similarity between user requirements and items in a dataset per particular attribute.

In the next step, fuzzy conformances are included into the aggregation operators to meet the user's preferences among attributes.

The simplest case for finding the best matching records is when a record is dominant by all atomic conditions, or is equal to all but one atomic condition and is better than the last one, i.e.

$$t_j v t_k \Leftrightarrow V_1(t_j) \geq V_1(t_k) \wedge \ldots \wedge V_{m-1}(t_j)$$
$$\geq V_{m-1}(t_k) \wedge V_m(t_j) > V_m(t_k) j, k = 1 \ldots n, j \neq k \quad (3)$$

where for the clarity, conformances are expressed as $C(A_i[t_u,t_j]) = V_i(t_j), i = 1 \ldots m$.

Often, a record can be more suitable by one and less suitable by another atomic conformance. This case is illustrated in Table 4 for conformances of seven attributes between user preferences t_u and records t_1 to t_5 in a dataset, where the values are taken from [16] for continuity with preliminary results and comparison of results.

The next section examines aggregation of atomic conformances to calculate overall matching degree for records in a dataset considering various kinds of preferences among attributes.

3 Aggregation of Atomic Conformances

This section examines several highly expectable cases of conformances' aggregations, which might be raised by users.

3.1 Aggregation of Equally Important Conformances

When all atomic conformances should be met at least partially, the solution is conjunction, generally expressed by t-norms.

The minimum t-norm adjusted for conformances (1) for record t_j is expressed as:

$$t_{min}(\mathbf{t_j}) = \min_{i=1,\ldots,n} C(A_i[t_u, t_j]) \tag{4}$$

where n is the number of atomic conditions. This t-norm provides the highest matching degree among all t-norms, but it lack the compensation effect [25, 26], i.e. higher conformances than the minimal one are simply ignored. The other t-norms have the property of downward reinforcement [25], e.g. the product t-norm, adjusted to conformances (1) as:

$$t_{prod}(\mathbf{t_j}) = \prod_{i=1}^{n} C(A_i[t_u, t_j]) \tag{5}$$

It produces the solution which is significantly lower than the lowest conformance, especially when we have higher number of atomic conformances.

Table 5 The aggregation of conformances from Table 4 by minimum and product t-norms, uninorm, and geometric mean

Record	Min (4)	Product (5)	Uninorm (6)	Geometric mean (7)
t_1	0.85	0.3771	1.00	0.8699
t_2	0.24	0.0006	1.00	0.3474
t_3	0.10	0.0047	1.00	0.4656
t_4	0.00	0.0000	0.00	0.0000
t_5	0.00	0.0000	0.00	0.0000

This problem is illustrated in Table 5 on the data from Table 4. When six attributes are conformant with value 0.85 each (record t_1) and one is congruent to value 1, the overall similarity to the user requirements is 0.3771 calculated by (5) (downward reinforcement). Users might conclude that this low value indicates that the considered flat is not very conformant with their requirements. It is not as highly rated as by minimum t-norm (4). But, the solution based on the minimum t-norm (4) copes with the problem of non-compensatory effect, ranking t_2 higher than t_3, even though t_3 is significantly more suitable in all but one conformance and worse in attribute A_1, because values higher than the minimum are not considered.

Disjunction is definitely not the option, because value 1 is annihilator. If a flat perfectly matches one conformance, the solution is 1 regardless the values of other conformances. Further, t-conorms, which model disjunction, also have one-directional, in this case the upward reinforcement property [25].

Therefore, an alternative may be hybrid function which covers conjunctive, disjunctive and averaging behaviour [27], because these functions meet the property of full reinforcement [25]. Such function attenuates low values (as conjunction does), emphasizes high values (as disjunction does) and calculates averages for low values of some conformances and high values for the reminder of the considered conformances. The 3-$\prod$ function [28] adjusted to calculate conformance (1) for record t_j is the following:

$$u_{3P}(\mathbf{t_j}) = \frac{\prod_{i=1}^{n} C(A_i[t_u, t_j])}{\prod_{i=1}^{n} C(A_i[t_u, t_j]) + \prod_{i=1}^{n}(1 - C(A_i[t_u, t_j]))} \tag{6}$$

The product in numerator (6) ensures that only the records (items) that at least partially meet all conditions are considered, i.e. value 0 is annihilator. Applying (6) on data in Table 4, has shown that $t_1 - t_3$ fully meet the condition whereas t_4 and t_5 are fully rejected. Record t_4 has conformance equal to 0 for attribute A_1 and therefore is excluded by both: t-norms and uninorms. But, the consequence of being mixed aggregation functions is that value 1 is also annihilator when other conformances are significantly met, which in our case makes indistinguishableness between items t_1, t_2, t_3 in Table 5.

Another option is averaging aggregation functions. The averaging function with the annihilator 0 is the suitable one to meet the requirement that all atomic conformances should be at least partially met. The averaging function with annihilator, among others, is geometric mean [29]. This function applied on conformances is as follows:

$$av_{geom}(\mathbf{t_j}) = \sqrt[1/n]{\prod_{i=1}^{n} C(A_i[t_u, t_j])} \tag{7}$$

Applying (7) on data in Table 5 (last column) has shown that t_1 is emphasized, but not as by uninorm (6), t_3 got better evaluation than t_2 as is expected due to better behavior in majority of conformances. Records t_4 and t_5 have conformances equal to 0 for one or more attributes and therefore are excluded by all considered functions: t-norms, uninorms and geometric mean. The geometric mean (7), as any averaging function gives solution which is between minimal and maximal value of conformances (except when minimal value is 0).

Although, t-norms are widely used in computing matching degrees for conjunction, the benefit of geometric mean should be used when a higher number of atomic conformances is aggregated. Further, the solution of geometric mean is between arithmetic mean minimum t-norm [25] causing that it is closer to conjunction than for instance quadratic mean. The drawback of considered uninorm (6) is that the value 1 is also annihilator (because of the disjunctive behavior in the disjunctive sub domain), see Table 5.

3.2 The Aggregation Considering Importance of Atomic Conformances and Their Groups

The main benefit of Choquet integral-based aggregation [30] is combining inputs, in our case conformances, when several atomic conformances create so-called coalitions (i.e. the contribution of one atomic condition is not important, but might be very important in the presence of high satisfaction degrees of some other atomic conditions).

An expression of the Choquet integral [25] adjusted to the conformance measures is as:

$$C_v(\mathbf{t_j}) = \sum_{i=1}^{n} [V_{(i)}(t_j) - V_{(i-1)}(t_j)] v(H_i) \qquad (8)$$

where for the clarity, conformances are expressed as $C(A_i[t_u, t_j]) = V_i(t_j)$ as in (3), $V_0(t_j) = 0$ by convention, $V_{(i)}(t_j)$ is a non-decreasing permutation of conformances for tuple t_j and v is a fuzzy measure of $Hi = \{(i), ..., (n)\}$. The fuzzy measure v is a set function [31]: $v : 2^{\mathcal{N}} \rightarrow [0, 1]$ which is monotonic and satisfies $v(\varnothing) = 0$ and $v(\mathcal{N}) = 1$, where $\mathcal{N} = \{1, 2, ..., n\}$.

For reasons of simplicity, we consider in the example attributes A_1 (distance to the grocery shop), A_2 (energy consumption), A_3 (size of flat) and A_7 (opinion from social networks). Attribute A_7 is less important, but when appears together with A_2 and A_3, its importance increases. The same holds for A_1 and A_3 (weight 0.40 each), but when both are considered, the weight is 0.70. The set function v is shown in Fig. 3.

Applying (8) on the data in Table 4 we conclude that the best tuple is t_1 followed by t_4 and t_5, see Table 6. The calculation for t_2 is ($C(A_2[t_u, t_2]) < C(A_3[t_u, t_2]) < C(A_7[t_u, t_2]) < C(A_1[t_u, t_2])$):

$$v(N)$$
$$1$$

$v(\{A_1, A_2, A_3\})$ $v(\{A_1, A_2, A_7\})$ $v(\{A_1, A_3, A_7\})$ $v(\{A_2, A_3, A_7\})$
     0.80   0.70   0.50   0.70

$v(\{A_1, A_2\})$ $v(\{A_1, A_3\})$ $v(\{A_1, A_7\})$ $v(\{A_2, A_3\})$ $v(\{A_2, A_7\})$ $v(\{A_3, A_7\})$
   0.60   0.70   0.30   0.45   0.30   0.30

$v(\{A_1\})$ $v(\{A_2\})$ $v(\{A_3\})$ $v(\{A_7\})$
     0.40   0.45   0.40   0.25

$$v(0)$$
$$0$$

Fig. 3 The set function v for attributes A_1, A_2, A_3 and A_7

	Record	Choquet (8)	Weighted arithmetic mean	Weighted geometric mean
Table 6 The aggregation of conformances from Table 4 by Choquet integral	t_1	0.8500	0.8483	0.8503
	t_2	0.3500	0.3220	0.3085
	t_3	0.4320	0.4170	0.3380
	t_4	0.6700	0.6601	0.0000
	t_5	0.6500	0.5720	0.4942

$$C_v(t_2) = 0.25v(N) + (0.26 - 0.25)v(\{A_1 + A_3 + A_7\})$$
$$+ (0.27 - 0.26)v(\{A_1 + A_7\}) + (0.5 - 0.27)v(\{A_1\}) = 0.35$$

The weighted arithmetic mean (when considering normalized weights of attributes depicted in Fig. 3) is 0.3220. The coalitions with A_7 are better than A_7 (the second best conformance, but low weight). Further, the group $\{A_1, A_3, A_7\}$ has higher weight than all individual attributes, hence the solution is greater than solution by the weighted mean functions. The same holds for tuple t_5. Although attribute of the highest weight has the lowest conformance, the solution is quite better due to influence of coalitions of attributes.

The Choquet integral is an averaging function, and therefore it is idempotent (t_1 in Table 6). Tuple t_4 is not excluded, because this structure of Choquet integral does not have annihilator. The important property of Choquet integral is possibility to represent interactions and dependencies among atomic conformances.

3.3 The Quantified Aggregation of Atomic Conformances

Regardless we use any t-norm, uninorm (6) or geometric mean (7), the solution is 0 when at least one atomic conformance is equal to 0. The option is applying a particular subset of averaging functions, e.g. Choquet integral, or quantified aggregation depending on the semantics of a considered problem.

When not all of the atomic predicates should be imperatively met we should consider the aggregation formalized by the fuzzy relative quantifier *most of* as: *most of atomic conditions should be met* [11] or, in our case, most of conformances should be greater than 0. For this purpose, we adjusted equation from the fuzzy quantified queries [15] to conformances in the following way:

$$qv(t_j) = \mu_Q(\frac{1}{n}\sum_{i=1}^{n}C(A_i[t_u, t_j])) \qquad (9)$$

where n is the number of conformances included into the query and μ_Q is the function of relative quantifier *most of* [32, 33] which can be re-formalized as:

$$\mu_Q(y) = \begin{cases} 0 & y \le 0.5 \\ \frac{y-0.5}{0.4} & y \in (0.5, 0.9) \\ 1 & y \ge 0.9 \end{cases} \qquad (10)$$

where y is the proportion in (9). Obviously, the ideal record is one with the conformance values equal to 1 for all attributes.

Regarding Tables 4 and 5, the best match is t_1 with validity 0.929, followed by t_4 with validity 0.743. Record t_3 has low validity, more precisely 0.157, and the validities of records t_2 and t_5 are zero. Although, record t_2 met all atomic preferences, low values of conformances are reflected in the proportion (9). On the other hand record t_4 failed to meet one conformance, but significantly met the other equally relevant ones. This aggregation is suitable when all atomic conditions are considered as soft ones, i.e. it is not an imperative that a particular atomic condition should be at least partially met, but majority of them should be significantly satisfied.

We should be careful, because this approach is not suitable when several conformances should be imperatively greater than zero (so-called hard conditions). For instance, when one of the considered attributes is price, i.e. the user cannot afford the flat that is beyond the budget, even when all other features are excellently conformant with the user's expectations. The next subsection suggests the solution for this case.

3.4 Merging Quantified Aggregation of Conformances with Hard Conditions

We should adjust quantified conditions to the cases when some of the atomic conditions must be met. For instance, if price is beyond the limit, it is irrelevant regardless all other conformances are perfectly match. Such conditions should be separately managed and aggregated with the rest of atomic conditions in the quantified condition (9) as:

$$m(t_j) = \left(\wedge_{i=1}^{p} C\left(A_i[t_u, t_j]\right)\right) \otimes \mu_Q\left(\frac{1}{q}\sum_{l=1}^{q} C\left(A_l[t_u, t_j]\right)\right) \qquad (11)$$

where p is the number of hard conditions, q is the number of atomic predicates (conformances) in the quantified aggregation of soft conditions and $\otimes$ is the binary aggregation function.

The same question regarding the suitable aggregation function holds in Sect. 3.1 and in (11). The suitable functions are conjunctions and geometric mean (7), because both sides should be satisfied. By the conjunctive aggregation function $\otimes$ in (11) we aggregate two predicates and therefore the reinforcement property is not so significant as for higher number of atomic predicates.

In Sect. 3.3, the second option is record t_4. However, if the conformance of attribute A_1 is a hard condition, e.g. instead walking distance it represents price, this record is irrelevant and therefore the aggregated value should be 0. The aggregation by (11) when $\otimes$ is replaced by minimum t-norm (4) is shown in Table 7. The results differ in comparison to previous sections, because the nature of preferences is changed.

For conjunction in (11), we can use any t-norm, but we should be aware of their strengths and weaknesses discussed in e.g., [25, 26] and illustrated in Sect. 3.1. The further evaluation has shown that geometric mean (7) and uninorm (6) are not suitable. For instance, if hard condition is satisfied with degree 0.1 and soft one with 0.7, the overall matching degree should not be higher than the degree of hard condition.

Table 7 The aggregation of hard condition and quantified condition for conformances in Table 4 by (11)

Record	hard condition (conformance of A_1)	quantified condition (conformances A_2 to A_7)	solution by min t-norm in (11)
t_1	0.850	0.875	0.850
t_2	0.500	0.000	0.000
t_3	0.100	0.350	0.100
t_4	0.000	1.000	0.000
t_5	1.000	0.000	0.000

4 Discussion

The inspiration for this work was problems, where a higher number of attributes influence the selection of the most suitable entities. But, collected values of considered attributes may be of the mixed data types, i.e. numerical, categorical or fuzzy. In addition, user may explain preferences for atomic condition by a data type which does not collide with the respective attribute in a dataset. Finally, user might raise a large scale of preferences among attributes.

Aggregation operators should be able to cover variety of preferences among atomic conditions, or in our case conformances. The conformance measure reveals how user requirements and items (records) in a dataset are conforming to the considered attributes regardless the non-uniformity in a data types. It can be straightforwardly used for manipulation with heterogeneous data types appearing in user's preferences and attributes' values. Further, managing conformances from the unit interval [0, 1] is convenient for the applicability of different aggregation functions in order to retrieve the most suitable records. Conformance measure provides flexibility to the users when considering similar items to the desired one or practitioners in the designing robust query engines.

When small number of atomic conditions is included where all of them should be at least partially met, the options are t-norms, which formalize conjunction in the fuzzy environment. But, when higher number of conformances is included, the averaging function like geometric mean is an option. When the best matches emphasized (upward reinforcement), and the weak matches attenuated (downward reinforcement); the solution can be reached by the uninorm function, e.g., 3-Π function (6). The product in nominator eliminates items, which fail to meet at least one conformance. But, when few conformances are fully met, item ideally meets requirements regardless other conformances (the disjunctive behavior on this part of domain).

The aggregation function, which meets the following requirements: 0 as annihilator, the compensation effect and value 1 as the neutral element, is the geometric mean. Further, the *ORNESS* measure [34] of this function is lower than 0.5 (i.e. for seven attributes it is 0.357), indicating that the solution is lower or equal than the one calculated by arithmetic mean (the *ORNESS* measure is always equal to 0.5 [25]) and therefore is closer to the solution by *MIN* function.

When a user provides a higher number of atomic conditions, where not all of them imperatively must be met, the aggregations by t-norms, uninorms and geometric mean are not suitable. The solution is aggregation by the relative quantifier *most of*, i.e. *most of conformances should be (significantly) met*. But, when one (or several) conformances must be met, the solution is aggregation between hard conditions (conformances which must be at least partially met) and soft conditions (it is beneficial if majority of these conformances are met) by t-norms or geometric mean.

Finally, when atomic conditions create so-called coalitions or relevant group, that is, a condition with low weight becomes very important in the presence of other atomic condition, aggregation by the discrete Choquet integral is the solution. This is an averaging aggregation, i.e. not excluding items which does not meet all atomic conditions. The Choquet integral-based approach is very useful when interaction and correlation between users' preferences appear in queries.

Nowadays, big data and their perspective for supporting informed decision making is in the focus of research. One of the big data features is variety of data types. This work has demonstrated suitability of conformance measures and aggregation functions for this feature research activities. The next feature is volume. Hence, the future research topic should be focused on examining efficiency and computing demands, as well as developing optimization procedures to improve efficiency.

5   Conclusions

In this work we have proposed method for querying heterogeneous data (different attributes expressed by diverse data types, or an attribute may convey diverse data types for different entities). Also, query conditions raised by users may be expressed by different data types than data types of respective attributes in a dataset. A robust mechanism requires flexible comparison measures for mixed data types. The fuzzy conformance has been proven to be a very useful approach. Fuzzy conformance is also the object of intense research activities in other fields such as discovering fuzzy functional dependencies, product recommendation techniques, data fusion in fuzzy relations etc.

The second part of this work was focused on the characterization of users' preferences based on different aggregation functions. This work provides more comprehensive analysis of applicability of aggregation operators in retrieving records from a datasets. Although t-norms are widely used in computing matching degrees of atomic conditions in queries, the benefit of geometric mean is obvious when higher number of atomic conformances is considered due to the compensatory effect and having value 0 as an annihilator.

When at least the majority of atomic conformances should be met, the solution is a quantified condition of the structure *most of atomic conformances should be met*. But, when several atomic conditions are hard (should be met), the solution is aggregation of hard and quantified condition by t-norms.

We should not neglect the cases, when weights of atomic conformances are not as important as weights of their particular groups, and not all atomic conformances should be imperatively satisfied. The solution for this task is Choquet integral-based aggregation.

Traditional datasets often contain data of the same type, but growing number of applications can deal with heterogeneous attributes. The large number of attributes having different data types is stored in many real datasets [35]. This study may help software developers to include further flexibility into the data queries, when mixed data types appear and users (e.g., decision makers and domain experts) provide higher number of desired features (atomic conditions). The overall matching degree is in the unit interval, which clearly indicates the distance of the considered records to the ideal or best one.

This field opens a space for the large scale of possible fusions of approaches dealing with databases, business intelligence, computational intelligence, data mining and decision making.

Acknowledgements This paper was partially supported by the project VEGA-1/0373/18 entitled "Big data analytics as a tool for increasing the competitiveness of enterprises and supporting informed decisions" by the Ministry of Education, Science, Research and Sport of the Slovak Republic

References

1. Snasel, V., Kromer, P., Musilek, P., Nyongesa, H. O., Husek, D.: Fuzzy Modeling of User Needs for Improvement of Web Search Queries. In: Annual Meeting of the North American Fuzzy Information Processing Society (NAFIPS 2007), pp. 446–451. IEEE, San Diego (2007).
2. Kacprzyk, J., Zadrożny, S.: FQUERY for Access: fuzzy querying for windows-based DBMS. In: Bosc, P., Kacprzyk, J. (eds.) Fuzziness in Database Management Systems, pp. 415–433. Physica-Verlag, Heidelberg (1995)
3. Wang, T-C., Lee, H-D., Chen, C-M.: Intelligent queries based on fuzzy set theory and SQL. In: 39th Joint Conference on Information Science, pp.1426–1432, Salt Lake City (2007).
4. Bosc, P., Pivert, O.: SQLf: a relational database language for fuzzy querying. IEEE Trans. Fuzzy Syst. **3**, 1–17 (1995)
5. Hudec, M.: An approach to fuzzy database querying, analysis and realisation. Computer Sciences and Information Systems **6**, 127–140 (2009)
6. Urrutia, A., Tineo, L., Gonzales, C.: FSQL and SQLf: Towards a standard in fuzzy databases. In: Galindo, J. (ed) Handbook of Research on Fuzzy Information Processing in Databases, pp. 270–298. Information Science Reference, Hershey (2008).
7. Škrbić, S., Racković, M: PFSQL: a fuzzy SQL language with priorities. In: 4th International Conference on Engineering Technologies, pp. 58–63. PSU-UNS, Novi Sad (2009).
8. Sözat, M., Yazici, A.: A complete axiomatization for fuzzy functional and multivalued dependencies in fuzzy database relations. Fuzzy Sets Syst. **117**(2), 161–181 (2001)
9. Sachar, H.: Theoretical aspects of design of and retrieval from similarity-based relational database systems. Ph.D. Dissertation, University of Texas at Arlington, Arlington (1986).
10. Vučetić, M.: Analysis of functional dependencies in relational databases using fuzzy logic. PhD thesis, University of Belgrade, Belgrade (2013).
11. Bosc, P., Hadjali, A., Pivert, O.: Empty versus overabundant answers to flexible relational queries. Fuzzy Sets Syst. **159**, 1450–1467 (2008)
12. Kacprzyk, J., Ziółkowski, A.: Database queries with fuzzy linguistic quantifiers. IEEE Transactions on Systems Man and Cybernetics SMC-16, 474–479 (1986).
13. Bosc, P., Brando, C., Hadjali, A., Jaudoin, H., Pivert, O.: Semantic proximity between queries and the empty answer problem. In: Joint 2009 IFSA-EUSFLAT Conference, pp. 259–264. EUSFLAT, Lisbon (2009).
14. Kacprzyk, J., Zadrożny, S.: Compound bipolar queries: combining bipolar queries and queries with fuzzy linguistic quantifiers. In: 8th Conference of the European Society for Fuzzy Logic and Technology (EUSFLAT 2013), pp. 848–855. EUSFLAT, Milano (2013).
15. Hudec, M.: Constraints and wishes in quantified queries merged by asymmetric conjunction. In: First International Conference Fuzzy Management Methods (ICFMsquare 2016), pp. 25–34. Springer, Fribourg (2017).
16. Vučetić M., Hudec M.: A Flexibile Approach to Matching User Preferences with Records in Datasets Based on the Conformance Measure and Aggregation Functions. In: 10th International Joint Conference on Computational Intelligence (IJCCI 2018), pp. 168–175. Seville (2018b).

17. Vucetic, M., Hudec, M., Vujošević, M.: A new method for computing fuzzy functional dependencies in relational database systems. Expert Syst. Appl. **40**, 2738–2745 (2013)
18. Vučetić, M., Hudec, M.: A fuzzy query engine for suggesting the products based on conformance and asymmetric conjunction. Expert Syst. Appl. **101**, 143–158 (2018)
19. Galindo, J.: Introduction and Trends to Fuzzy Logic and Fuzzy Databases. In: Galindo, J. (ed.) Handbook of Research on Fuzzy Information Processing in Databases, pp. 1–33. Information Science Reference, Hershey (2008)
20. Zadeh, L.A.: Fuzzy sets as a basis for a theory of possibility. Fuzzy Sets Syst. **1**, 3–28 (1978)
21. Shenoi, S., Melton, A.: Proximity relations in the fuzzy relational database model. Fuzzy Sets Syst. **100**, 51–62 (1999)
22. Tung A.K.H., Zhang, R., Koudas, N., Ooi, B.C.: Similarity search: a matching based approach. In: 32nd International Conference on Very Large Data Bases, pp.751–762. Seoul (2006).
23. De Pessemier, T., Dooms, S., Martens, L.: Comparison of group recommendation algorithms. Multimedia Tools and Applications **72**(3), 2497–2541 (2014)
24. Herrera, F., Martínez, L.: A model based on linguistic 2-tuples for dealing with multigranular hierarchical linguistic contexts in multi-expert decision-making. IEEE Transactions on Systems, Man, and Cybernetics Part B **31**, 227–234 (2001)
25. Beliakov, G., Pradera, A., Calvo. T.: Aggregation Functions: A Guide for Practitioners. Springer-Verlag, Berlin Heidelberg (2007).
26. Klement, E.P., Mesiar, R., Pap, E.: Triangular Norms. Kluwer, Dordrecht (2000)
27. Dubois, D., Prade, H.: On the use of aggregation operations in information fusion processes. Fuzzy Sets Syst. **142**, 143–161 (2004)
28. Yager, R., Rybalov, A.: Uninorm aggregation operators. Fuzzy Sets Syst. **80**, 111–120 (1996)
29. Beliakov, G., Bustince Sola, H., Calvo Sánchez, T.: A Practical Guide to Averaging Functions. Springer, Berlin (2016)
30. Choquet, G.: Theory of capacities. Ann. Inst. Fourier 5, (1954).
31. Wang, Z., Klir, G.: Fuzzy Measure Theory. Plenum Press, New York (1992)
32. Kacprzyk, J., Yager, R.R.: Linguistic Summaries of Data Using Fuzzy Logic. Int. J. Gen Syst **30**, 133–154 (2001)
33. Zadeh, L.A.: A computational approach to fuzzy quantifiers in natural languages. Comput. Math. Appl. **9**, 149–184 (1983)
34. Dujmović, J.J.: Two integrals related to means. Publikacije Elektrotehničkog fakulteta. Serija Matematika i fizika. 232–236 (1973).
35. Bashon, Y., Neagu, D., Ridley, M.J.: A framework for comparing heterogeneous objects: on the similarity measurements for fuzzy, numerical and categorical attributes. Soft. Comput. **17**(9), 1595–1615 (2013)

Reconciling Adapted Psychological Profiling with the New European Data Protection Legislation

Keeley Crockett⬤, Jonathan Stoklas, James O'Shea, Tina Krügel, and Wasiq Khan⬤

Abstract Adaptive Psychological Profiling systems use artificial intelligence algorithms to analyze a person's non-verbal behavior in order to determine a specific mental state such as deception. One such system known as, Silent Talker, combines image processing and artificial neural networks to classify multiple non-verbal signals mainly from the face during a verbal exchange i.e. interview, to produce an accurate and comprehensive time-based profile of a subject's psychological state. Artificial neural networks are typically black-box algorithms; hence, it is difficult to understand how the classification of a person's behaviour is obtained. The new European Data Protection Legislation (GDPR), states that individuals who are automatically profiled, have the right to an explanation of how the "machine" reached its decision and receive meaningful information on the logic involved in how that decision was reached. This is practically difficult from a technical perspective, whereas from a legal point of view, it remains unclear whether this is sufficient to safeguard the data subject's rights. This chapter is an extended version of a previous published paper in IJCCI 2019 [35] which examines the new European Data Protection Legislation and how it impacts on an application of psychological profiling within an Automated Deception Detection System (ADDS) which is one component of a smart border control system known as iBorderCtrl. ADDS detects deception through an avatar border guard interview, during a participants' pre-registration, to demonstrate the challenges faced in trying to obtain explainable decisions from models derived

K. Crockett (✉) · J. O'Shea · W. Khan
Department of Computing and Mathematics, Manchester Metropolitan University, Manchester, M1 5GD, UK
e-mail: k.crockett@mmu.ac.uk

J. O'Shea
e-mail: j.d.oshea@mmu.ac.uk

J. Stoklas · T. Krügel
Institute for Legal Informatics, Leibniz Universität Hannove, Hannover, Germany
e-mail: jonathan.stoklas@iri.uni-hannover.de

T. Krügel
e-mail: kruegel@iri.uni-hannover.de

© The Author(s), under exclusive license to Springer Nature Switzerland AG 2021
C. Sabourin et al. (eds.), *Computational Intelligence*, Studies in Computational Intelligence 893, https://doi.org/10.1007/978-3-030-64731-5_2

through computational intelligence techniques. The chapter concludes by examining the future of explainable decision making through proposing a new Hierarchy of Explainability and Empowerment that allows information and decision-making complexity to be explained at different levels depending on a person's abilities.

Keywords Psychological profiling · Deception detection · Artificial neural networks · Decision trees · GDPR

1   Introduction

Psychological Profiling is a technique best known as a tool used within criminal investigations utilising methodologies from both law enforcement and psychology [1]. It involves the detailed and intricate analyses of the non-verbal behaviour of a person, often in an interview situation, to detect their mental state. The expertise and training required by a human to undertake this kind of profiling is complex—requiring simultaneous conjecture of many non-verbal signals. Adaptive psychological profiling utilises computational intelligence techniques to build models of non-verbal behaviour for different mental states, i.e. deceptive behaviour or more recently to detect comprehension levels in education. For example, Silent Talker (ST) [2], a profiling system for lie detection, uses hierarchies of neural networks to module deceptive behaviour. However, neural networks are by nature 'black boxes' in which it is difficult to understand how the trained networks determine if a person is deceiving or not.

The European data protection reform package that is applicable since May 2018 consists of the General Data Protection Regulation (Regulation 679/2016/EU, "GDPR") and the "Law Enforcement Directive (680/2016/EU). The GDPR potentially has a worldwide impact on business models and research activities carried out within industry and academic institutions that utilise computational intelligence (CI) algorithms with respect to personal data [3]. Specifically, it states the rights of an individual not to be subject to automated decision-making, such as profiling, unless it is (a) necessary for entering into or performance of a contract between the data subject and a data controller, (b) authorised by Member State's laws, or (c) it is based on the explicit consent of the data subject. In addition, in any aspect of automated decision-making, the individual has the right to human intervention (opt out) as well as to be provided with an explanation of how the automated decision was reached. This would be achieved through disclosure of "the logic involved" (article 13, GDPR, [4]). When profiling, the data controller should use appropriate mathematical and statistical procedures and data should be accurate (up to date and free from bias) in order to minimize the risk of errors.

This legislation presents many challenges when using CI for modelling complex problems that involve people. How do we provide an explainable decision suitable for all stakeholders when using 'black box' CI algorithms? The stakeholders are the experts who designed, validated and tested the system, the business or customer who commission the system and the data subject who receives the automated decision from the system. This chapter explores this issue using an application of an automated deception detection system (ADDS) utilised within a pilot system known as iBorderCtrl, which detects deception through an avatar border guard interview during a participant's pre-registration. The final stage of the deception detection architecture is a single neural network classifier. In Sect. 5 experiments, this is re-placed with a traditional decision tree to provide a set of rules on how decisions about deceptive behaviour are reached. The complexity and size of the rule sets produced show, that whilst an expert, may have some understanding of the rules, it would be extremely difficult for a member of the public to understand and even the expert could at present not be able to say precisely why these particular rules were derived or explain what they mean.

Section 2 of this chapter defines what is meant by psychological profiling in the context of this work, whilst Sect. 3 explains the legal perspective of some aspects of automated decision making in light of the GDPR. The case study of profiling EU travellers is described in Sect. 4 and used to illustrate the challenges of developing explainable profiling systems. Section 5 provides the methodology used to conduct empirical experiments on a deception detection profiling system and presents results using both neural networks and decision trees in terms of explainability. This section also considers the future of explainable decision making in terms of a proposed Hierarchy of Explainability and Empowerment. Finally, Sect. 6 provides some important considerations for both the legal and computational intelligence communities.

2 Psychological Profiling

Adaptive Psychological Profiling is the process of determining a person's internal mental state (beliefs, desires, and intentions) through analysing their external behaviour by means of Computational Intelligence (CI) based components. Furthermore, it is based on a generic architecture, which is adapted to different application domains and optimised through a process of machine-learning. The first such architecture is known as "Silent Talker" (ST) [2, 5]. ST uses complex interactions between non-verbal features in a moving video feed from an interviewee to classify truthful or deceptive behaviour. The ST architecture has been adapted for different internal mental states. One such adaptation is for comprehension in intelligent tutoring, in the classroom [6]. Another ethnic/cultural adaptation extended comprehension classification to Tanzanian women for informed consent in a clinical trial [7]. Other ongoing work includes an avatar based deception detection interview integrated into a smart border crossing system [8, 9]. The case study used in this chapter focuses on the complex problem of the psychological profiling of deceivers—the next section looks more deeply into the science of lying and why this in particular is a challenging problem for computational intelligence in terms of building a model and in trying to explain automated decision making.

2.1 The Science of Lying

There are various different types of lie, with different contextual motivations and different ways of classifying them. For example, Ganis et al. [10] used two classes, whether the lies fit into a coherent story and whether they were previously memorized. Alternatively, Feldman et al. [11] presented a taxonomy of lies with 10 codings, for lies produced by participants with three different self-presentational goals. Regardless of context, there is a general psychological principle that the act of deceiving produces changes in behaviour which has a long history dating back to the Hindu Dharmasastra of Gautama, (900–600 BC) and the philosopher Diogenes (412–323 BC) according to Trovillo [12].

There are a number of factors, proposed by psychologists, which may be influential drivers of behavioural change during deception. These include general arousal/stress, cognitive load, behaviour control and special cases of arousal, guilty knowledge and duping delight. Stress is the oldest driver to be measured for lie detection. Following work by Angelo Mosso in the late 19th and early twentieth centuries using pulse and blood pressure, the polygraph was invented by Larson in 1921 (International League of Polygraph Examiners, [13]). The Cognitive Load driver derives from the work of George A. Miller [14], whose Magical Number 7 (± 2) indicated that there were a limited number of "mental variables" that an individual could process concurrently. Therefore, someone trying to construct and remain consistent with a false account would be under increased cognitive load. Behaviour control occurs when deceptive interviewees deliberately try to control themselves in order to make an honest and convincing impression. It is postulated that attempts to control behaviour will increase in higher-stakes scenarios [15]. Guilty knowledge (Concealed Knowledge) is a test of whether a suspect has information related to a crime that an innocent person would not possess. When exposed to such information an interviewee is expected to produce a reaction detected by instrumentation [16]. Duping delight is believed to occur in an interview when the deceiver experiences pleasurable excitement at the prospect of successfully deceiving the interviewer, particularly in the presence of observers [17].

2.2 Automated Lie Detection

The field of computational intelligence provides a wealth of algorithms which are suitable to build models of deceivers automatically. Silent Talker (ST) [2, 5 and 18] differs from many other lie detectors in its assumption that deceptive non-verbal behaviour is the outcome of a combination of psychological drivers and that it cannot be characterized by a simple, single indicator. ST uses complex interactions between multiple channels of microgestures over time to determine whether the behaviour is truthful or deceptive. A microgesture is a very fine-grained non-verbal gesture, such as the right-eye moving from half-open to closed. This can combine with other micro-gestures from the right eye to detect a wink or both eyes to detect a blink. Measured over time these can combine to measure blink rate. Complex combinations and inter-actions of (typically) 38 channels and interactions between them can be compiled into a long vector, over a time slot, which can be used to classify behaviour as truthful or deceptive over the slot. Microgestures are significantly different from micro expres-sions (proposed in other systems), because they much more fine-grained and require no functional psychological model of why the behaviour has taken place. Further-more, because there are so many channels contributing to the analysis, behaviour control is infeasible. Typically, using a recording device such as a web cam, salient features (e.g. an eye) are identified in an individual video frame by a layer of object locators. The states of the objects are detected by the pattern detectors (e.g. eye half open). The channel coders compiled the outputs of the pattern detectors over time (e.g. sequence of eye movement indicating a blink) and the deception classifier uses this long vector compiled by the channel coders. The ST approach to lie detection is based on a "black box" model, the conjecture that these and other (unknown) factors act as drivers of non-verbal behaviour, resulting in distinctive features that can be used to discriminate between deceivers and truth-tellers. Silent Talker is in itself an automated profiling system and is being piloted as a basis for an automated deception detection system to profile travellers crossing European borders at a pre-registration phase and will be described in Sect. 4 of this chapter.

3 The GDPR

3.1 Automated Decision-Making Under the GDPR

From a legal perspective, various issues arise. In 2016, the European Union agreed on a data protection reform package including the General Data Protection Regu-lation (GDPR) (Regulation (EU) 2016/679, [4]), which is applicable since 25 May 2018. The GDPR introduces various new regulations which affect both profiling and the use of computational intelligence-based systems. Despite the fact that all general provisions have to be met, such as the data protection principles (art. 5 GDPR) and procession upon a legal basis (art. 6 GDPR), as explained above, for profiling and automated decision-making there is a specific provision in art. 22 GDPR. According to art. 22 (1) GDPR, an automated decision is a decision based solely on automated processing, including profiling of a person, which produces legal effects concerning him or her or similarly significantly affects him or her. One of the most obvious chal-lenges in this regard is the fact that almost any decision in an increasingly digitized world might have at least a mediate legal effect as well [19]. Therefore, the inter-pretation of this requirement should be rather restrictive [20], whereas any single case shall be assessed based on objective factors [22]. While this result seems to be sound and necessary, persons without a legal background might face difficulties in assessing whether their decision is to be seen as automated decision-making or not. An automated lie detection system, however, will most probably always cause significant effects on the persons, both deriving from the possible use-cases, as well as with regard to personality rights in general (e.g. reputation).

3.2   Safeguards for Automated Decision-Making

For decisions falling within the scope of art. 22 GDPR, certain safeguards need to be considered. Following the principles described in art. 5 GDPR, automated decision-making only can take place for specific purposes, and it must be necessary, legitimate and proportionate. Also, according to art. 22 (3) GDPR, the data controller shall *"implement suitable measures to safeguard the data subject's rights and freedoms and legitimate interests, at least the right to obtain human intervention on the part of the controller, to express his or her point of view and to contest the decision."*

Consequently, it is not possible to apply automated decision-making without offering a secondary system which involves human beings. In that regard, automated decision-making would act as a filter, where only those cases which would result in negative consequences for the data subject would be checked by a human being. However, it is important to ensure that such a check by human beings is not biased by the previous automated results. From a legal point of view, it remains unclear whether decision assistance systems are being covered by art. 22 GDPR. According to the wording of art. 22 GDPR, it shall only apply to a decision based solely on automated processing, which does not include assistance systems [21]. In fact, art. 22 GDPR only ensures that human beings are not subject to a decision by a machine [22]. Therefore, at present assistance systems are in general not covered by the specific provisions on automated decision-making included in the GDPR. However, to qualify as human involvement, it therefore has to be ensured, that the involvement is carried out by someone who has both, the authority and competence to change the decision [20]. Nevertheless, it might be possible that human beings, even though they have the authority and competence will just follow the system's recommendation, in particular if the accuracy of such a system is extremely high. Proper training and sensitization (both with regard to the risks deriving from automated decision-making and knowledge on fundamental rights) is required. Also, organizational measures to ensure that human beings checking automated decisions do not have any negative consequences, such as the burden of proof or liability when deviating from the machine's decision, are required, as this would otherwise undermine the right to human intervention and would lead to a de-facto decision through the machine.

Apart from these issues raised by the safeguards explicitly mentioned in the GDPR, further issues can arise from the principle of non-discrimination. While art. 22 GDPR aims to ensure that the data subject does not face any negative consequences through a decision made by a machine, specific provisions on bias in algorithms cannot be found in the GDPR. Instead, general principles such as data accuracy (art.

5 lit. d) GDPR) and fundamental rights need to be considered. In the context of auto-mated decision-making and fundamental rights, bias of algorithms is one of the most relevant issues. While from a technical point of view, a decision would be only wrong if the system was not working as intended, results can, although technically correct, nevertheless be wrong from an ethical and legal point of view. Biased algorithms, therefore, are rather a regulatory issue than a technical one. If an algorithm relies on biased data in any of the steps for development and use, results will be biased as well [23]. A common ground for this can be found in both disciplines, though: Data used for machine-learning and automated decision-making needs to be accurate, of good quality and shall not lead to discrimination. Therefore, on the input side, preventing and detecting inaccurate data is crucial for the application of artificial intelligence and machine-learning. For prevention, an impact assessment and the selection of data to learn from is a very important safeguard to ensure that the data subject's funda-mental rights are not being violated; for the ongoing use on the output side, tools to constantly monitor decisions and identifying possible issues with bias is required [23]. As one of the most important steps to avoid bias is transparency [24], the "black box" phenomenon as described above is a specific challenge in that regard.

For the use of artificial intelligence in general, but also in the context of deception detection and border security, further issues can be identified, such as a violation of human dignity (i.e. caused by an increase in human–machine interaction with smart systems), for instance due to an objectification of the human being and a possible stigmatisation caused by false positives. It has to be noted, though, that these issues rather derive from other fundamental rights than the right to privacy, meaning that the GDPR alone is not sufficient to tackle the challenges for the legal framework raised by artificial intelligence.

Consequently, additional legislation might be required. However, whether these changes have to be addressed by a (fundamental) reform or rather lead to small adjustments depending on the use-cases in which artificial intelligence is applied [25], is still under discussion. Legislative approaches such as the strategy on artifi-cial intelligence in Germany, which contains the establishment of a legal and ethical framework for research on and the use of AI [26], or the European parliament resolu-tion containing recommendations to the Commission on Civil Law Rules on Robotics [27], or the Draft Ethics Guidelines for trustworthy AI [28], show that the legislator is increasingly aware of the issues, while precise regulations are not available yet. Also, the principle of proportionality requires all legislative approaches to consider if new technologies are actually beneficial for the society. This needs to be followed by an assessment on how ethical issues could be mitigated and the rights and freedoms of citizens can be safeguarded.

3.3 Information Obligations and Their Extent

Additional obligations can be found in the data subject's rights: According to art. 13 (2) lit. (f), 14 (2) lit. (g) and 15 (1) lit. (h) GDPR, the data controller is required to inform the data subject about:

- the existence of automated decision-making as referred to in art. 22,
- meaningful information about the logic involved, and
- the significance and the envisaged consequences of such processing for the data subject.
- In addition, information has to be provided in concise, transparent, intelligible and easily accessible form, Art. 12 (1) GDPR. This also applies to information obligations regarding automated decision-making [28].

These regulations, however, are not sufficiently clear from a legal point of view [29]. While it shouldn't be a practical problem to inform about the existence of automated decision-making, it remains unclear what is meant with "meaningful information on the logic involved". According to the German Federal Court of Justice, an algorithm can be a trade secret, but the data subject has the right to be informed about which personal data is being used to compute a decision [30]. It has to be noted though, that the case was decided in 2013, meaning that the GDPR has not been considered. Also, the ruling did not consider computational intelligence based systems and the "black box" phenomenon, but an algorithm used for credit scoring. Therefore, the question could be raised as to whether the principles of the ruling could still apply considering the impact of computational intelligence and the GDPR's transparency requirements. For further interpretation of the requirement to provide "meaningful information on the logic involved", recital 63 of the GDPR could be used: It states that the right to access to personal data should not "adversely affect the rights or freedoms of others, including trade secrets or intellectual property and in particular the copyright protecting the software." This would, on the one hand, imply that the protection of trade secrets as decided by the German Federal Court of Justice is still a valid argument. On the other hand, it has to be noted that computational intelligence has the potential to hugely affect daily life. Transparency for such decisions is crucial to ensure that people are not stigmatised or discriminated against.

However, comparing automated decision-making to human decision-making—which can, (at present) to an even bigger extent, also affect daily life—it becomes obvious, that human decision-making also lacks transparency. Despite the fact that already from a medical point of view the (intentional or unintentional) true motivation of a human being is not comprehensible, there is no right to such transparency towards human decision-making processes. What we have in different cases is the right to an (ex post) explanation (e.g. court or recruiting decisions). Such obligations, therefore, only refer to the result and not the decision-making process, and do not ensure that a human being—even if he/she was obliged to be transparent—would tell the truth, leaving a risk that a decision would not be transparent but only justifiable. Some analogy could be drawn to the distinction between the "context of discovery" and the "context of justification" as it is discussed in scientific theory [31]. Considering that CI-based systems are increasingly capable of competing with human beings in terms of their ability to interpret information, the question could be raised why computer systems have to be fully transparent, when in fact the principle of full transparency is not applied for interactions between human beings. However, the principle of human dignity ensures that human beings do not have to reveal their inner thoughts, whereas such a right does not exist for machines. While of course a certain degree of transparency is necessary to ensure that a system is not discriminating against people or otherwise violating laws, it will have to be discussed as to how transparent a computer system has to be.

Besides this, information does not necessarily lead to more transparency. Quite the contrary, extensive information often overburdens the person concerned. Therefore art. 12 (1) GDPR states that information must be provided in an intelligible way. Considering the increasing complexity of algorithms and machine learning approaches, this means that it might not be possible to reveal the technical functioning of an algorithm in an intelligible way, but only a simplified description, e.g. on which data is being used and how. This general information most probably is not sufficient to adversely affect trade secrets.

However, with regard to the specific use of deception detection for security purposes such as border control, revealing any information on the functioning of an algorithm, including the categories of personal data, which have been processed, might reveal confidential information about the procedures of security agencies. In these cases, information to be provided might be further restricted and other measures have to be implemented, such as expert groups or ethics commissions.

As shown, the GDPR is only partially capable of addressing the aforementioned issues and it is questionable whether it would be the right place to address these issues. Having a specific regulation on algorithms might be beneficial both for users and end-users of computational intelligence based systems and crucial to guarantee that the rights and freedoms of persons affected are being respected.

3.4 Intelligible Information for the Data Subject

Another legal issue with regard to the information obligations is the requirement to provide intelligible information. This leaves room for various interpretations: Should the information be intelligible for the data controller, for the individual data subject, or rather for an objective, reasonable and informed third party? [32]. In that regard, it needs to be considered that data controllers might have a substantial advantage both in knowledge of their systems and technology in general. While detailed technical information could be on the one hand seen as a maximum level of transparency, an average data subject will most probably not be able to understand such information. Therefore, information should be less detailed than it would be theoretically possible to provide, if this ensures that the data subject can actually understand the information. This is also reflected in art. 12 and recital 58, stating that the data subject shall be addressed using clear and plain language. However, it needs to be ensured that such simplified information is sufficient to enable the data subject to understand the impact of this decision on his fundamental rights, including the right to informational self-determination and appointed expert groups capable to verify more detailed information could be introduced to control the reasoning on a more detailed level.

Last but not least, another challenge is the fact that data subjects can have very different background-knowledge helping them to understand information. People who frequently use ICT services might be more familiar with the functioning of algorithms than people who can barely use a computer. If the GDPR is required to ensure that every *individual* data subject understands the logic involved, data controllers would have no legal certainty as to whether they comply with the legal requirements or not. Therefore, the information provided to describe the logic involved should be intelligible for an objective, reasonable and informed third party persons ("the average user"), while at the same time providing as much information as possible.

3.5 Challenges for the Technical Community

As outlined above, many issues regarding automated decision-making derive from legal obligations. However, there are certain issues which also require input and solutions from the technical community, in particular:

- How to properly assess the legal situation regarding automated decision-making and how to apply proper safeguards?
- How to explain an algorithm without leaking trade secrets?
- How can algorithms based on computational intelligence be explained?
- Can the information on how an algorithm learns be sufficient to understand it's functioning and decision-making?
- Can self-learning algorithms also explain their decision-making, and could this be updated frequently for every user?

4 Case Study: Profiling Traveller's Across Schengen Borders

iBorderCtrl, (Intelligent Portable Control System) is a three year H2020 research and innovation project, funded by the European Union, which is developing novel mobility concepts for land border security. The system will enable authorities to achieve higher throughput at the crossing points through faster processing of passengers within vehicles or pedestrians whilst guaranteeing high security levels through targeting criminal activities such as human trafficking, smuggling and terrorism. In addition, the system will aim to reduce the subjective control and workload of human border agents and to increase the objective control through non-invasive automated technologies. The aim is to ensure that travellers have a speedier border crossing through engaging in a pre-registration step [8]. A full description of the project can be found here: https://www.icross-project.eu/. iBorderCtrl features a unique combination of state-of-the-art biometric tools which will provide risk scores to a Risk Based Assessment Tool (RBAT) that will act as an automated decision-maker on the status of the traveller as they arrive at the border crossing point (Green is proceed, Amber is second line check and Red is recommended refusal—a refusal recommendation would require multiple sources of evidence including document checks, biometric checks etc.). It is important to say that iBorderCtrl is a human in the loop system and therefore provides advice to human border guards who ultimately have the final say. The focus in this chapter is on the profiling of travellers deceptive behaviour in the pre-registration step using a psychological profiling system called ADDS (Automated Deception Detection system) [33] and how such a system when deployed in the field, provides numerous challenges if asked to provide an explainable decision to different stakeholders: the research and development team of ADDS, the Border Guards and their managers and the travellers using the system. The next section provides an overview of ADDS.

4.1 Automated Deception Detection System

In the pre-registration phase, after entering the information about their trip each traveller will be required to be interviewed by a Border Guard avatar. Information is exchanged between the iBorderCtrl System and the ADDS system using a unique QR code which is generated for each trip a traveller goes on. An overview of the ADDS components can be found in Fig. 1. At the start of the interview, the traveller will be provided with instructions on how to align themselves with the camera on their own device. They will then start the interview. The system will present the traveller with one question at a time. After each spoken response, the traveller will be asked to confirm their answer using radio button responses. This acts as a timestamp between questions and responses and is recorded by ADDS. After each question, analysis on deceptive behaviour is undertaken and at the end of the interview, an overall deceptive risk score is calculated. Then this is uploaded to the cloud based iBorderCtrl database where it will make a weighted contribution (based on the ADDS technical readiness level) to the overall risk profile of a traveller.

In the pilot research studies, consenting adults who meet the ethical criteria and agree to take part, will be asked 16 questions, similar to those asked at border crossing points. For example: What will your means of travel into the Schengen Area be? What is the purpose of your trip? The interview will last on average 2.5 min subject to network connections and speed. ADDS will be responsible for conducting the interview where the Avatar asks questions, utilising three attitudes (puzzled, neutral and positive) and two avatars, one male and one female which in the first testing phase

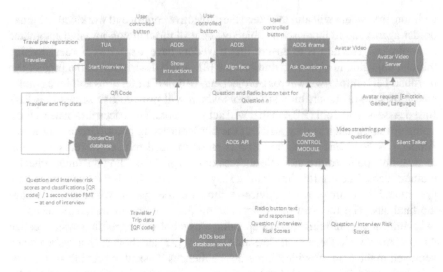

Fig. 1 ADDs component overview [34]

will be randomly assigned. An example of a female Border Guard avatar (designed by Stremble Ventures [35] is shown in Fig. 2. For the purpose of this research and the pilot study, the avatar speaks the question verbally. The traveller then confirms their verbal answer with one extracted from information they have already provided in relation to this question or they can ask the avatar to repeat the question (Fig. 2).

Fig. 2 Female avatar border guard [34]

Fig. 3 Backend silent talker processing [34]

The non-verbal behaviour of each video question response will be analysed by the Silent Talker system [33] which will output the deceptive risk score for each question. ADDS utilises 38 non-verbal channels which vary in complexity. Each channel is coded to the bipolar measurement range [-1, 1] by the Channel Accumulator [33] and these are ultimately grouped into channel vectors based upon a time slot (i.e. 1 to 3 s) before being encoded within the image vector. Figure 3, shows an example of the backend processing carried out by the Silent Talker component of ADDS. The bottom screen, displays the live video stream for a specific interview question, whilst on the right the number of truthful, deceptive and unknown slots are shown. At the conclusion of the interview, questions and interview risk scores and their associated classifications, along with one second's worth of video frames are uploaded to the iBorderCtrl database to be used by Face Matching Module [36] and RBAT [9]. This prevents attempts to cheat the system by impersonation as the Face Matching Module will check that the traveller who performs the interview is the same as the person on their identity document. Further confirmation of the identity will take place when the traveller reaches the border. In summary, the fundamental level of truthful or deceptive classification is based on a timeslot, a snapshot of non-verbal behaviour during an interview. Timeslots are aggregated over a complete answer to a question to classify the behaviour over the complete answer. Either timeslots or answer classifications may be aggregated over the whole interview to give an overall classification for the interview.

5   Methodology

Initial experiments were conducted based on a typical airport security scenario of packing a suitcase in order to train, validate and test ADDS for use as a module component in the iBorderCtrl system. An empirical study was conducted using 30 participants whose non-verbal behaviour was recorded whilst engaged in an online interview with a static border guard avatar [33]. The aim of the experiment was to derive models of truthful and deceptive non-verbal behaviour using a configuration of artificial neural networks and traditional decision trees to classify deception and truthfulness.

The hypothesis tested was:

H0: A decision made by an automated deception profiling system cannot be explained using decision tree models.

H1: A decision made by an automated deception profiling system can be explained using decision tree models.

Each participant first read a participant information sheet and had the ability to ask questions to the researchers, before signing an informed consent form. The participants then took part in a role play exercise which was designed in a similar manner to the suitcase scenario reported in [33]. The activity first involved packing a suitcase with items typically taken on a holiday. Participants were also asked to look at posters typically found at airports showing prohibited items when boarding an aircraft. Each participant was randomly assigned either a truthful or one of four deceptive scenarios designed to cover high and low stakes deception. For example, based on the literature, it was anticipated that a person transporting drugs would generate higher arousal levels than a person being deceptive about packing some illegal agricultural produce. Following the role play, participants were then interviewed by a border guard avatar and were asked 13 questions which are typical of those asked by border guards. This is a different, scenario specific, set of questions from the generic set described for pilot testing in Sect. 4.1. Full details of the experimental methodology are described in [33]. Following data preparation (described in Sect. 2.1), two classification models were developed. One based on the hierarchal ANN model used by Silent Talker and the other using decision tree rule induction which aims to construct a set of rules which will classify objects from knowledge of a set of examples whose classes are previously known. The method is based on recursive partitioning of the sample space and defines classes structurally by using decision trees. The well-known benefit of decision trees is their ability to provide transparency of the decision-making process. For the purpose of this work, Quinlan's C4.5 decision tree algorithm were used [37].

5.1   Data

86,584 image vectors were collected from the image data of the 30 participants, where each vector contained the states of each of the 38 non-verbal channels. Ground truth was established for each participant's interview question through knowledge of the scenario that they role-played. I.e. truthful (43,051 image vectors) or deceptive (43,535 image vectors). Out of the 32 participants, there were 17 deceptive and 15 truthful interviews, 22 males and 10 females with a mix of ethnicities. For the purpose of the ADDS experimental scenario reported in this chapter, the final ANN classifies truthfulness or deceptiveness is based upon an activation level in the range [−1, 1] which was determined from the data set. The deception risk score, Dq, of each of the questions was defined as

$$D_q = \frac{\sum_{s=1}^{n} d_s}{n} \qquad (1)$$

where d_s is the deception score of slot s and n is the total number of slots for the current question. In order to obtain a classification for each vector, the following thresholds were applied.

IF Question_risk (D_q) < = x THEN.
Image vector class = truthful (-1).
ELSE IF Question_risk (D_q) > = y THEN
Image vector class = deceptive (+1).
ELSE
Indicates not classified.
END IF

Where x = −0.05 and y = +0.05. Thus if a question risk score was within this range, a classification could not be allocated. This range is empirically defined, determined through previous work and only used for the initial experimental scenarios described in [33].

5.2 Results and Analysis

In [33], two methods for training, validation and testing were reported: *n-fold* cross validation and leave a pair out. The latter being a more appropriate measurement of accuracy for unseen participants, which is required when a system such as ADDS is deployed in the field. However, for the purposes of this chapter, in the context of comparing models in terms of classification accuracy and their ability to produce an explainable decision, cross validation is used, as initial work showed there was little difference in induced decision tree size. With no ANN or C4.5 optimisation, the best tree from performing *tenfold* cross validation contained 1072 rules. Table 1 shows the overall classification accuracy of tenfold cross validation for both the ANN (ADDS-ANN) and C4.5 (ADDS-DT). The additional results of the probabilistic classifier Naïve Bayes are also shown for comparative purposes.

```
lhleft   <= -0.407407
|       lright <= 0.777778
|   |     fmuor   <= 0.072831
|   |   |     rhright <= -0.809524
|   |   |   |     fbla   <= -0.07892
|   |   |   |   |     fblu   <= -0.773891
|   |   |   |   |   |     rleft   <= 0.793103
|   |   |   |   |   |   |     rright   <= 0.310345
|   |   |   |   |   |   |   |     rhclosed <= -0.933333
|   |   |   |   |   |   |   |   |     fhs <= -0.888889
|   |   |   |   |   |   |   |   |   |     fmuor   <= 0.028317
|   |   |   ·|   |   |   |   |   |   |   |     lright <= -1
|   |   |   |   |   |   |   |   |   |   |     rleft    <= -1
|   |   |   |   |   |   |   |   |   |   |   |     fbla    <= -0.997762
|   |   |   |   |   |   |   |   |   |   |   |   |     fblu   <= -0.963101
|   |   |   |   |   |   |   |   |   |   |   |   |   |     fmc   <= -0.942354: TRUTH (10.0)
|   |   |   |   |   |   |   |   |   |   |   |   |   |     fmc   > -0.942354: DECEPTION (9.0)
|   |   |   |   |   |   |   |   |   |   |   |   |     fblu > -0.963101
|   |   |   |   |   |   |   |   |   |   |   |   |   |     lhleft   <= -1: DECEPTION (180.0)
|   |   |   |   |   |   |   |   |   |   |   |   |   |     lhleft   > -1
|   |   |   |   |   |   |   |   |   |   |   |   |   |   |     Ethnicity <= -1: TRUTH (10.0)
|   |   |   |   |   |   |   |   |   |   |   |   |   |   |     Ethnicity > -1: DECEPTION (27.0)
|   |   |   |   |   |   |   |   |   |   |   |   |     fbla    > -0.997762
|   |   |   |   |   |   |   |   |   |   |   |   |     Ethnicity <= -1
|   |   |   |   |   |   |   |   |   |   |   |   |   |     fmuor   <= -0.030638
|   |   |   |   |   |   |   |   |   |   |   |   |   |   |     fhs <= -1: DECEPTION (59.0)
|   |   |   |   |   |   |   |   |   |   |   |   |   |   |     fhs > -1: TRUTH (18.0)
|   |   |   |   |   |   |   |   |   |   |   |   |   |     fmuor   > -0.030638
|   |   |   |   |   |   |   |   |   |   |   |   |   |   |     lclosed   <= -0.034483: TRUTH (140.0)
|   |   |   |   |   |   |   |   |   |   |   |   |   |   |     lclosed   > -0.034483
|   |   |   |   |   |   |   |   |   |   |   |   |   |   |   |     flm   <= -0.9: TRUTH (29.0)
|   |   |   |   |   |   |   |   |   |   |   |   |   |   |   |     flm   > -0.9: DECEPTION (6.0)
|   |   |   |   |   |   |   |   |   |   |   |   |     Ethnicity > -1: DECEPTION (29.0)
|   |   |   |   |   |   |   |   |   |   |   |     rleft   > -1: DECEPTION (127.0)
|   |   |   |   |   |   |   |   |   |   |     lright > -1: DECEPTION (521.0/2.0)
|   |   |   |   |   |   |   |   |   |     fmuor   > 0.028317
|   |   |   |   |   |   |   |   |   |     Gender <= -1: DECEPTION (17.0)
|   |   |   |   |   |   |   |   |   |     Gender > -1
|   |   |   |   |   |   |   |   |   |     lright <= -1
```

Fig. 4 Rule snapshot [34]

Are Decisions Explainable? Fig. 4. shows a snapshot of the best decision tree which contained 1072 rules and had a tree size of 2143 nodes.

The rules induced from the dataset represent patterns of non-verbal behaviour for specified channels which, when combined, allow the classification of deception verses truth for a given risk score. One rule from this tree which gives a classification of deception can be extracted as follows:

IF lhleft < -0.407407 AND lright < = 0.777778 AND fmuor < = 0.072831 AND rhright < = 0.310345 AND rhclosed < = -0.93333 AND fhs < = -0.888889 AND fmour < = 0.028317 and lright < = -1 and rleft < = -1 and fbla < = -0.997762 and fblu < = -0.963101 and fmc > -0.942354 THEN CLASS DECEPTION.

Note that this is a relatively simple rule as it deals with summary statistics for each channel. Analysing the rule, one familiar with the underpinning theory of Silent Talker will see information on four non-verbal channels associated with the eyes: left eye looking left (lleft), left eye looking right (lright), right eye half closed (rhclosed), right eye looking left (rleft) and 5 channels containing information about the state of the face including the horizontal movement of the face (fhs) face angular movement up-on-right (fmuor) and the degree of blushing/blanching (fblu). Face channels track the face movement along the X-axis and Y-axis using the coordinates and dimensions of the face found by the Face Object Locator ANN [7]. Likewise, the state of each eye channel is determined from a Pattern Detector ANN [33] observing the left/right eye image and/or from the application of logical decision(s). The values for each channel are determined empirically by the pattern detector ANNS and the channel encoder ANNS in the bipolar range [1 and -1].

In this application, the rules are complex and look at combinations of fine grained non-verbal behaviour i.e. movement of facial features. Due to this complexity, individual rules are difficult for an average human to comprehend. They could not for example be replicated by a human. As the problem is complex, the tree is large— previous work [38] suggests pruning may lead up to a 25% reduction in rules. A sacrifice in classification accuracy occurs but still the quantity of rules is large and difficult to comprehend. But is this problem scenario based? If automated profiling were applied to a simpler more typical problem, such as a bank loan or mortgage application then perhaps the learned rules could be understood by all stakeholders—the expert, the member of the public and the bank manager. Consider for example, a small dataset containing 434 instances for applications for personal loans. 238 instances are reject samples and 196 accept. The dataset contains just 14 attributes. Using C4.5 and *tenfold* cross validation, a classification accuracy of 74.8% is achieved and the best tree contains 27 rules. A sample rule is shown below:

IF TimeAtBank(years) < = 2 AND TimeEmployed(years) < 1 AND ResidentStatus = "Rented" THEN Outcome = REJECT LOAN.

A person, profiled by this system, could have the decision explained to them using this rule by a staff member at the bank i.e. they had not been a customer at the bank for long enough and had not been in employment for over a year and they currently lived in rented accommodation. What the staff could not do is qualify the accuracy of the model used to train the system, explain and show the statistical evidence behind the decision, nor guarantee that there was any bias in the training data that led to the model. Therefore, neither the hypothesis H0 nor H1 can be universally accepted as explainability is determined by problem representation and complexity.

5.3   The Future of Explainable Decision Making

Section 5.1 provides evidence that, at least at the most detailed technical level, some Computational Intelligence (CI) decision making will be completely inexplicable even to a reasonably well-informed and educated public. Still from a legal point of view, there is no distinction in the GDPR with regard to the information obligations as to how complex a decision is. However, this should not provide reasons for abandoning the use of CI or the duty of explainability. We propose a progressive

Increasing
Mathematical
Complexity

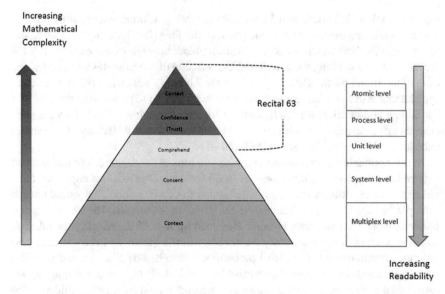

Increasing
Readability

Fig. 5 O'Shea's hierarchy of explainability and empowerment

approach to explaining decisions and challenging them, based on a hierarchy of capability. Understanding information at a particular level of complexity will result in a particular capability to question and object—and data subjects should have a right to question, challenge and receive answers appropriate to their level of attainment. The proposed Hierarchy of Explainability and Empowerment is shown in Fig. 5.

There is a clear disjunction in the hierarchy. One might question whether the top three levels can be intelligible at all, whereas reaching the consent level should be considered as the minimum requirement for a CI system making decisions automatically. At present, the best approach to ensure transparency seems to be to maximise readability of text explanations and minimise the mathematical complexity of formulae in explanations to data subjects. Recital 63, also covering trade secrets or intellectual property (see Sect. 3.3) may prevent a full technical disclosure pertinent to the upper levels of the hierarchy. However, taking into account the abovementioned arguments to what extent the right to information shall be interpreted, an answer could be the following: From a technical point of view, there is the possibility of creating a simpler abstraction, although in the area of "know-how" (such as data cleansing) even procedures that appear quite simple may need to be protected for commercial reasons. On the other hand, anything which is patented is automatically fully-disclosed (see Sect. 5.2.5). Each level will now be explained.

The Context Level. This level considers the Computational Intelligence systems as component of a multiplex (a collection of systems), which itself should be contestable at a societal level. The multiplex should be understandable and the overall process of decision making should be challengeable by data subjects as citizens, through (usually elected) representatives with a duty, time and resources to be better informed than the general public and legislate on their behalf for example, see [39]. In one sense, the actions at this level are the drivers or enablers for the Consent level. As the multiplex level may contain multiple black boxes, it is legitimate for the public to express themselves in terms of subjective feelings and decisions at this level. Decision-making will largely rely on verbal reasoning and debate. Because the particular CI component is part of an overall multiplex which may contain other CI and Information Systems, the data subject should be entitled to know which components were influential in making the decision and how their contributions were combined. In particular, the subjects should be informed whether the CI system was influential in the decision taken about them and if so, which of the other systems in the multiplex confirmed or contradicted the CI system. At this level, it should be established that the data subject should be allowed to contest the decision without having to make a technical argument.

The Consent Level. This level concerns the ability to understand the personal consequences of using the CI system, to give consent or opt out and to object to an automated decision. As this is crucial to allow for a lawful processing on common legal bases such as consent (art. 6 (1) lit. a. GDPR) or legitimate interests (art. 6 (1) lit. f. GDPR), this level should be seen as the minimum requirement regarding information provided to the data subject. It should provide information about the basic quantitative measures used by the system. For example, the percentage accuracy in making decisions with contextual information about humans making the same decisions.

An example invitation to do the pre-travel interview in the border control context might be:

"I am inviting you to do a pre-travel interview. This could make your border crossing faster.

A Computational Intelligence system will watch your interview and calculate the risk that you were lying.

The system calculates other risks, for example from a machine reading your passport.

The lie detector is not 100% accurate, we know this and allow for it when we calculate risks.

No lie detection method, including human experts, is 100% accurate.

The current accuracy of the system is 75%.[1]

If the combined risks are high, you will have an interview with a border guard to clear things up.

Would you like to take the pre-travel interview?".

This passage has Flesch Readability Ease score of 60.9 and a Flesch-Kincaid Grade Level of 7.8, and is described as "Plain English. Easily understood by 13- to 15-year-old students." (USA).

An example notification of the right to contest in the border control domain might be a statement such as:

"Your answers in the pre-travel interview showed there was a risk that you weren't telling the truth.

---

[1] We expect to achieve 85% accuracy if such a system were deployed.

> The passport scanner had difficulty reading your passport.
>
> The border control system thinks there may be a risk if allows you to cross the border.
>
> Machines can make mistakes. The current accuracy of the deception detection system is 75%.
>
> This is your chance to clear things up by speaking to a human border guard."

This passage has Flesch Readability Ease score of 74.8 and a Flesch-Kincaid Grade Level of 5.8, and is described as "Fairly easy to read." In the border control example, the text should be accompanied by a contingency table (not using technical jargon such as "false alarm"), for example,

> These are the results of our tests of the deception detector component. You can see how well it performs with truthful people and deceptive people that it has never seen before. Errors are shown in red. These errors show you more about how the CI system could have made a mistake in interviewing you and may help you decide whether or not you wish to contest the decision.

Truthful cases	Deceptive cases	
75.55	24.55	Classified as truthful
26.34	73.66	Classified as deceptive

A contingency table helps the data subject to understand whether the CI classifier is biased towards one particular outcome or not. It is worth noting, that the results shown in the contingency table are from preliminary experiments conducted in 2017 on 30 people and published in [33]. Understanding the contingency table requires numeracy skills of understanding and reasoning with percentages and proportions, corresponding to upper key stage 2 of the UK national curriculum, year 6 (aged 10 and above).

This level cannot explain the logic of a CI system as a detailed algorithm, but it does explain how the system reasoned about its calculations and relates this reasoning to the inputs, for data subjects with a general education. While this would probably allow the data subject to better decide on whether he/she would like to ask for human intervention, it would probably not provide meaningful information on the logic involved in the algorithm, as required by the GDPR.

The Comprehend Level. This level concerns the ability to integrate understanding of various aspects of the CI system to envisage their consequences for decision making. It involves some disclosure of the logic involved in the automated decision-making, but involves a tradeoff between transparency & accessibility on one hand and providing meaningful information on the other.

This requires drilling down into the black box system to reveal it at unit level (as far as possible). It supports understanding the relationships between components (units) working together in the CI system. Data subjects should be able to compose a more informed question or objection about the correct performance of a component in the overall system.

Numeracy skills will become more important at the comprehend level, however it should be possible to provide meaningful quantitative descriptions within the scope of upper key stage 4 of the UK national curriculum (years 10 &11, aged 14–16),

Considering the pre-travel interview for border control, the data subject may be informed of the roles of components, their inputs and outputs, and how the data flows between them to make an overall classification, provided there is no conflict with Recital 63 IP Protection. Concepts such as signed numbers, thresholding and inequality relations characterize reasoning about these components.

An *example notification* of the right to contest in the border control domain might be a passage such as:

The Computational Intelligence system breaks each question up into a series of short video clips (slots).

Each video clip is processed by the system and given a risk score between -1 (minus one) and $+1$ *(plus one)*.

-1 shows a strong belief that you were truthful.

+1 shows a strong belief that you were deceptive.

Numbers in between show different strengths of belief.

The CI system then checked if the risk score was less than -0.05. If it was, it classified your video clip as truthful.

Otherwise, it tested if your answer was greater than $+0.05$. If it was, it classified your slot as deceptive.

For risks between -0.05 (very weak truthful) and $+0.05$ (very weak deceptive) it decided it was not confident enough to classify the slot.

The lie detector is not 100% accurate and some answers will be classified incorrectly, but it ignores slots that contain weak indicators and it calculates the score for the complete answer by comparing the number of deceptive video clips to the total number of video clips for an answer.

The classifications for your answers were:

1. *What is Your Surname?—Deceptive*

(etc.)

Machines can make mistakes. You may have good reasons and evidence to challenge these classifications.

This is your chance to clear things up by speaking to a human border guard.

This passage has Flesch Readability Ease score of 63.8 and a Flesch-Kincaid Grade Level of 7.6, and is described as "Plain English. Easily understood by 13- to 15-year-old students." The numeracy requirements to understand and challenge at this level include concepts such as signed numbers, thresholding and inequality relations (Sect. 5.1). This correspond to upper key stage 4 of the United Kingdom national curriculum, years 10 &11 (aged 14–16).

It should be noted that, in the examples in this section, the interview answers given by the data subject have been classified (to some degree) as deceptive. Formally, this does not preclude a data subject from challenging a favourable (i.e. truthful) classification by the system.

The Confidence Level. This level discloses information about processes within the CI system. At this level, there is more likely to be conflict between protecting the Intellectual Property in the CI system and informing the data subject. The intention is to demonstrate that their data is represented accurately, that appropriate procedures are applied and that consequently the risk of a profiling error has been minimised. Data representation is, superficially, the simplest element. The data used to make the decision can be disclosed for the subject to check allowing them to be sure that they have not been misidentified or to provide evidence to correct errors of fact in records. Some data may not be disclosable for reasons of security—however, this is not a particular issue for a CI system since the same would apply if the data were processed by human border guards (for example, information in a law enforcement database obtained through intelligence gathering). However, the data belonging to a data subject undergoes mathematical transformations when processed by a CI system, some may be explainable at this level, others may only be explainable at the contest level.

Some performance figures have already been disclosed at the lower levels of the hierarchy. Understanding can be enhanced at the confidence level with statistical analysis—not only did the CI component perform with a certain accuracy during testing, but this is the level of confidence that the figure quoted did not just occur by chance.

A further disclosure is the demographic composition of data sets used to train and test the system currently deployed. This will allow data subjects to challenge the result of the test based on lack of representation of their gender, ethnicity or other possible confounding factors. Data subjects viewing the system at this level will need to understand concepts such as statistical significance, α and p values, population samples, frequency distributions etc. to make effective use of the information provided. The mathematical skills required correspond to AQA A-level statistics in the UK (aged 16–18).

An *example notification* of the right to contest in the border control domain might be a passage such as:

> The Computational Intelligence system analysed the video of your answers to questions by locating non-verbal features (mainly facial features), detecting their states and combining these, over time, to feed a final classifier.
>
> In this system all of the tasks were performed by Artificial Neural Networks.
>
> ANNs are created using sets of training and testing data, from which they learn solutions that apply to cases they have never seen.
>
> Therefore, they require representative data sets.
>
> The data sets used in iBorderCtrl were composed of (percentages for gender distribution, percentages for ethnic distribution).
>
> The ANNS trained for locating and detecting states of features achieved at least x% accuracy.

The ANN for finally classifying deception achieved 75% accuracy.

The p-values the ANNs detecting and locating were all less than the α value of 0.05, the normal acceptable level for scientific tests. Some were less than α value of 0.01 which is the stronger level for scientific testing.

The p-value for the ANN finally classifying deception was n.nn which is less than the α value of p.pp etc.

Machines can make mistakes. You may have good reasons and evidence to challenge these classifications, this is your chance to clear things up by speaking to a human border guard.

This passage has Flesch Readability Ease score of 49.9 and a Flesch-Kincaid Grade Level of 10.8, and is described as "Difficult to read" and "Sophomore" level in the education system (USA). The numeracy requirements to understand and challenge at this level include concepts such as statistical significance, α and p values, population samples, frequency distributions etc. to make effective use of the information provided. The mathematical skills required correspond to AQA A-level statistics in the UK (aged 16–18).

The Contest Level. The Contest level discloses information at the atomic level, allowing a sufficiently skilled data subject with state-of-the-art understanding of the CI system to contest its decisions in detail. This level of disclosure assumes that there is no conflict with Recital 63. It should be noted that Recital 63 will not always be an obstruction. Gaining patent protection for commercial use of such a system requires sufficient information to be disclosed for "one sufficiently skilled in the art" to reproduce it.

At this level, the data subject should be able to express a fully-informed point of view backed by evidence at the level of algorithm functioning. Data subjects will need advanced understanding of CI and the domain, for example topics such as calculus, entropy and transfer functions. This would typically require education to honours degree level (including specialist CI units of study) or PhD level. Data subjects with this level of skill and education would be informed through peer-reviewed scientific publications and patents. Theoretically, this level would enable the data subjects to track their data through the system, replication of the mathematical operations, checking their results and tracing the steps to the final classification. This would have little or no value to the typical data subject due to the infeasibility in terms of time for the human to replicate the computer calculations on their data and the opaqueness of the classification (we know what happened, but we don't know why it happened).The real value of this level of disclosure is in the protection offered by expert scrutiny and validation of the system by independent scientists.

Concluding Meaningful versus Intelligible Information. The various levels as described above outline the differences in the level of detail which could be applied when informing a data subject on the logic involved when processing personal data, including automated decision-making. As described in Sect. 3.4, this information, however, also must be intelligible. While transparency is crucial to enable the data subject to conclude informed decisions, a high level of detail, which the data subject might not be able to understand, does in fact not increase transparency, but rather is contraindicative. Similar to legislation on consumer protection, information obligations should focus on being short and straight to the point, while including all relevant information.

In general, the consent level can be seen as minimum requirement for any information obligation deriving from a data processing operation. In most use-cases relating to GDPR, the data subject needs to be enabled to make informed decisions. However, for certain use-cases such as for border checks or other security related matters, revealing information on the logic involved might need to be limited quite exhaustively, even to an extent below the consent level. While in that case the context level might be still sufficient to generally inform the data subject on the fact that automated decision-making is being used, additional safeguards would be needed. Such a data processing operation would in any case require a statutory legal basis, but also additional safeguards such as auditing algorithms and oversight mechanisms to ensure that, even though the data subject's rights are restricted, the risk of violations of fundamental rights is minimised.

Regarding uses-cases without such restrictions, it needs to be monitored how the understanding and education of average users will develop in the future. While for the most users intelligible information nowadays might be covered by the consent or comprehend level, this might change over time, requiring to also include information from the comprehend, confidence and contest level. Consequently, the concept of information obligations has to be understood as a flexible model, allowing for adjustments and refinements whenever required.

Depending on the impact and relevance on the data subject, certain use-cases might also require an additional monitoring of algorithms used for automated decision-making through specialised entities such as NGOs, expert groups or oversight authorities. While such experts could also understand the top-levels of the proposed model, the risk of infringements of the data controller's interests could be minimised, in particular regarding IP protection, while at the same time ensuring that an algorithm does not violate the data subject's fundamental rights without him/her being able to actually realise this.

6   Conclusions

This chapter has used a case study of adaptive psychological profiling to examine the challenges of how to produce explainable decisions of CI models to all stakeholders. There are many challenges for both the computational intelligence and the legal communities. Therefore, finding solutions that reflect technical realities while at the same time providing sufficient privacy safeguards is crucial. This, however, requires a close collaboration of both the legal and technical community, which currently happens very rarely. A closer collaboration between both communities would allow better guidance, such as common guidelines for software developers, standardized frameworks which comply with the GDPR by default, and many more [40]. Therefore, receiving answers to the questions raised in Sect. 3.5 would be an important first

Table 1 Tenfold cross validation results

Method	%Train AVG	%Test AVG	%Class-Accuracy
ADDS-ANN	97.03	96.66	96.80
ADDS-DT	98.90	98.80	98.80
ADDS-Naïve Bayes	80.12	70.12	75.12

step to further deepen the common understanding of technical and legal challenges relating to the GDPR and to foster a debate on the proper interpretation of the GDPR among the legal community, as well as information on how the technical community could be supported in their efforts to comply with legal requirements. Finally, the future of explainable decision making is expressed as a new proposed Hierarchy of Explainability and Empowerment which allows information and decision making complexity to be explained at different levels depending on the abilities of a person. Examples are given of the hierarchy for responding to different needs of explainability. Such a hierarchy can be seen as the conceptual starting point in addressing explainability in CI systems.

Acknowledgements The iBorderCtrl project has received funding from the European Union's Horizon 2020 research and innovation programme under grant agreement No 700626. Silent Talker research has been supported by Silent Talker Ltd.

References

1. Bonn, S.: Criminal profiling: the original mind hunters, profiling killers dates back to jack the rip-per. In psychology Today (2017). [online] https://www.psychologytoday.com/ us/blog/wicked-deeds/201712/criminal-profiling-the-original-mind-hunters [Accessed 27/7/2018]
2. Silent Talker Ltd.: (2019) [online] https://www.silent-talker.com/. [Accessed 13/01/2019]
3. Bryan Casey, B., Farhangi, A., Vogl, R.: Rethinking explainable machines: the GDPR's "right to explanation" debate and the rise of algorithmic audits in enterprise. (2018) [online] https://papers.ssrn.com/sol3/papers.cfm?abstract_id=3143325 [Accessed 21/01/2019]
4. GDPR Portal: (2018) [online]. Available at: https://www.eugdpr.org/ [Accessed 27/07/2018]
5. Bandar, Z., Rothwell, J., McLean, D., O'Shea, D.: Analysis of the behaviour of a subject, Publication date 2004/5/3, p. 10475922. Patent office US, Application number (2004)
6. Holmes, M., Latham, A., Crockett, K., O'Shea, J.: Near real-time comprehension classification with artificial neural networks: decoding e-Learner non-verbal behavior. In IEEE Transactions on Learning Technologies, Issue: 99 (2017). DOI: https://doi.org/10.1109/TLT.2017.2754497
7. Buckingham, F., Crockett, K., Bandar, Z., O'Shea, J., MacQueen, K., Chen, M.: Measuring human comprehension from nonverbal behaviour using artificial neural networks. In: Proceedings of IEEE World Congress on Computational Intelligence Australia, pp. 368–375 (2012), DOI: https://doi.org/10.1109/IJCNN.2012.6252414
8. Crockett, K.A., O'Shea, J., Szekely, Z., Malamou, A., Boultadakis, G., Zoltan, S.: Do Europe's borders need multi-faceted biometric protection? In Biometric Technology Today. **7**, 5–8 (2017). ISSN 0969–4765

9. iBorderCtrl Intelligent Portable Control System [online]. Available at https://www.iborderct rl.eu/ [Accessed 07/01/2019]
10. Ganis, G., Kosslyn, S.M., Stose, S., Thompson, W.L., Yurgelun-Todd, D.A.: Neural correlates of different types of deception: an fMRI investigation. Cereb. Cortex **13**(8), 830–836 (2003)
11. Feldman, R.S., Forrest, J.A., Happ, B.R.: Self-presentation and verbal deception: do self-presenters lie more? Basic Appl. Soc. Psychol. **24**(2), 163–170 (2002)
12. Trovillo, P. 1939. History of Lie Detection, In Journal of Crim. L. & Criminology, 848.
13. International League of Polygraph Examiners: Polygraph/Lie Detector FAQs (2016). [Online] https://www.theilpe.com/faq_eng.html. [Accessed 28/7/2018]
14. Miller, G.: The magical number seven, plus or minus two: some limits on our capacity for processing information **63**(2), 81–97 (1956)
15. Caso, L., Gnisci, A., Vrij, A., Mann, S.: Processes underlying deception: an empirical analysis of truth and lies when manipulating the stakes. J. Invest. Psychol. Offend. Prof. **2**(3), 195–202 (2005)
16. MacLaren, V.: A quantitative review of the guilty knowledge test. J. Appl. Psychol. **86**(4), 674–683 (2001, Aug)
17. Sen, T., Hasan, M.K., Tran, M., Yang, Y., Hoque, M.E.: Say Cheese: common human emotional expression set encoder and its application to analyze deceptive communication. In Automatic Face & Gesture Recognition (FG 2018), 2018 13th IEEE International Conference on (pp. 357–364) (2018, May)
18. Rothwell, J., Bandar, Z., O'Shea, J., McLean, D.: Silent talker: a new computer-based system for the analysis of facial cues to deception. J. Appl. Cogn. Psychol. **20**(6), 757–777 (2006)
19. Von Lewinski: In BeckOK DatenschutzR DS-GVO Art. 22, Fn. 28. (2018) [online] https://beck-online.beck.de/?vpath=bibdata%2fkomm%2fBeckOKDatenS_26%2fEWG_DSGVO%2fcont%2fBECKOKDATENS%2eEWG_DSGVO%2eA22%2eglB%2eglIV%2egl1%2ehtm. [Accessed 27/07/2018]
20. Art. 29 Data Protection Working Party: WP251rev.01, Guidelines on Automated individual decision-making and Profiling for the purposes of Regulation 2016/679, p. 21 f. (2018) [online] https://ec.europa.eu/newsroom/article29/document.cfm?doc_id=49826 [Accessed 21/01/2019]
21. Von Lewinski: In BeckOK DatenschutzR DS-GVO Art. 22, Fn. 2. (2018) [online] https://beck-online.beck.de/?vpath=bibdata%2fkomm%2fBeckOKDatenS_26%2fEWG_DSGVO%2fcont%2fBECKOKDATENS%2eEWG_DSGVO%2eA22%2eglA%2ehtm. [Accessed 27/07/2018]
22. Martini: In Pal/Pauly DS-GVO 2nd Edition 2018, Art. 22, Fn. 20a (2018). [Accessed 27/07/2018]
23. Fundamental Rights Agency: Preventing unlawful profiling today and in the future: a guide, p. 12 (2018a). [Online] https://fra.europa.eu/en/publication/2018/prevent-unlawful-profiling. [Accessed 17/01/2019]
24. Fundamental Rights Agency: Preventing unlawful profiling today and in the future: a guide, p. 60 (2018b). [Online] https://fra.europa.eu/en/publication/2018/prevent-unlawful-profiling. [Accessed 17/01/2019]
25. Borges, G.: Rechtliche Rahmenbedingungen für autonome Systeme, NJW 2018, 977 (979) (2018)
26. Bundesregierung (German federal government): Klarheit und Kontrolle bei Algorithmen 2018. [online] https://www.bundesregierung.de/Content/DE/Artikel/2017/07/2017-03-07-maas-algorithmen.html [Accessed 26/11/2018]
27. European Parliament.: European Parliament resolution of 16 February 2017 with recommendations to the Commission on Civil Law Rules on Robotics (2015/2103(INL)) (2017) [online] https://www.europarl.europa.eu/sides/getDoc.do?type=TA&language=EN&reference=P8TA-2017-0051 [Accessed 26/11/2018]
28. The European Commission's high-level Expert Group on Artificial Intelligence: Draft Ethics Guidelines for Trustworthy AI (2018). [online] https://ec.europa.eu/newsroom/dae/document.cfm?doc_id=56433 [Accessed 21/01/2019]

29. Martini.: In Paal/Pauly DS-GVO 2nd Edition 2018, Art. 22, Fn. 41a (2018c)
30. Bräutigam/Schmidt-Wudy: Das geplante Auskunfts- und Herausgaberecht des Betroffenen nach Art. 15 der EU-Datenschutzgrundverordnung, CR 2015, 56 (62) (2015)
31. German Federal Court of Justice: (2014). VI ZR 156/13. [online] https://juris.bun desgerichtshof.de/cgi-bin/rechtsprechung/document.py?Gericht=bgh& Art=en&nr=66910& pos=0&anz=1 [Accessed 21/01/2019]
32. Schickore, J., Steinle, F.: Revisiting discovery and justification. Histor. Philo-Soph. Perspect. Cont. Distin. (2006). https://doi.org/10.1007/1-4020-4251-5
33. Schmidt-Wudy.: In BeckOK DatenschutzR DS-GVO Art. 15, Fn. 78. (2018b) [online] https:// beck-online.beck.de/Dokument?vpath=bibdata%2Fkomm%2Fbeckokdatens_24%2Fewg_ dsgvo%2Fcont%2Fbeckokdatens.ewg_dsgvo.a28.htm. [Accessed 27/07/2018]
34. O'Shea, J., Crockett, K., Khan, Kindynis, P., Antoniades, A.: Intelligent deception detection through machine based interviewing. In Proceedings of IEEE International Joint conference on Artificial Neural Networks (IJCNN), DOI: https://doi.org/10.1109/IJCNN.2018.8489392
35. Crockett, K., Stoklas, J., O'Shea, J., Krügel, T., Khan, W.: Adapted psychological profiling verses the right to an explainable decision, 10th International Joint Conference on Computational Intelligence, Seville, Spain, (2018). ISBN: 978–989–758–327–8
36. Stremble Ventures.: (2019) [online] https://stremble.com/. [Accessed 28/11/2019]
37. Rodriguez, L., Hupoint, I., Teno, C.: Facial recognition application for border control. Proceedings of IEEE International Joint Conference on Artificial Neural Networks (IJCNN) (2018). https://doi.org/10.1109/IJCNN.2018.8489113
38. Quinlan, R.: C4.5: Programs for Machine Learning. Morgan Kaufmann Publishers (1994). ISBN 1–55860–238–0
39. O'Shea, J., Crockett, K., Khan, W.: A hybrid model combining neural networks and decision tree for comprehension detection. Proceedings of IEEE International Joint Conference on Artificial Neural Networks (IJCNN) (2018). https://doi.org/10.1109/IJCNN.2018.8489621
40. O'Shea, J.: Dr James O'Shea—written evidence (to the UK House of Lords Artificial Intelligence Select committee), AIC0226, published 26 October 2017, https://www.parliament.uk/ business/committees/committees-a-z/lords-select/ai-committee/publications/
41. Crockett, K., Goltz, S., Garratt, M.: GDPR impact on computational intelligence research. Proceedings of IEEE International Joint Conference on Artificial Neural Networks (IJCNN) (2018). https://doi.org/10.1109/IJCNN.2018.8489614

Concordance in *FAST-GDM* Problems: Comparing Theoretical Indices with Perceived Levels

Marcelo Loor, Ana Tapia-Rosero, and Guy De Tré

Abstract A key task while trying to reach consensus within a *flexible attribute-set group decision-making* (FAST-GDM) problem is the quantification of the level of concordance between the evaluations given by each participant and the collective evaluations computed for the group. To cope with that task, several theoretical concordance indices based on similarity measures designed to compare intuitionistic fuzzy sets (IFSs) have been introduced. Also, a visual tool, called *IFS contrasting chart* (IFSCC), has been proposed to facilitate an estimation of the level of concordance between individual and collective evaluations characterized respectively by two IFSs. Aiming to determine the level to which those theoretical indices reflect the perceived levels of concordance, in this paper we introduce a novel variant of an IFSCC, which includes an explicit ranking that provides additional information regarding a potential decision on the evaluated options. We use this variant in a test where individuals were asked to estimate the level of concordance between collective and individual evaluations obtained while solving a FAST-GDM problem. The results of this test and some suggestions about the use of such theoretical indices are presented.

This paper is an extended version of the work published in: Loor, M., Tapia-Rosero, A., De Tré, G.: *Usability of Concordance Indices in FAST-GDM Problems*. In: Proceedings of the 10th International Joint Conference on Computational Intelligence (IJCCI 2018). pp. 67–78. INSTICC, SciTePress (2018). DOI: 10.5220/0006956500670078.

M. Loor · G. De Tré
Dept. of Telecommunications and Information Processing, Ghent University,
Sint-Pietersnieuwstraat 41, 9000 Ghent, Belgium
e-mail: Guy.DeTre@UGent.be

M. Loor (✉) · A. Tapia-Rosero
Dept. of Electrical and Computer Engineering, ESPOL Polytechnic University, Campus Gustavo Galindo V., Km. 30.5 Via Perimetral, Guayaquil, Ecuador
e-mail: Marcelo.Loor@UGent.be

A. Tapia-Rosero
e-mail: atapia@espol.edu.ec

Keywords Flexible consensus reaching · Group decision-making · Intuitionistic fuzzy sets · IFS contrasting charts

1 Introduction

Group decision-making is usually understood as a process where a group of participants try to make a collective decision about what is the best option to take [4]. For instance, consider a situation in which a group of chocolate makers are trying to make a decision about the best variety of cacao beans for making chocolate bars from a new collection of cacao beans grown in South America. In this case, the chocolate makers can reach a collective decision about the best variety(ies) of cacao beans for making chocolate bars by going through a process in which, under the supervision of a moderator, they iteratively reconsider their evaluations to be in agreement with the group. In such a process, the moderator can assume that the chocolate makers have a similar expertise and, thus, ask them to evaluate each variety using a predefined collection of attributes that are inherent to any of the varieties under evaluation. This situation, as so described, constitutes an example of a *multi-attribute group decision-making* (MA-GDM) problem [5, 11].

A contrasting situation is one in which participants with different expertise hold mixed opinions on which attributes should be taken into account to evaluate the given options. As an example, consider another situation where three friends, say Alice, Bob and Chloe, who are not trained chocolate makers, have been asked to make a collective decision about the best variety(ies) for making chocolate bars from the new collection of cacao beans: while Alice considers that '*Criollo*' variety is almost the best for making chocolate bars because it is delicate, Bob considers this variety to be unacceptable for making chocolate bars because it is rare and very expensive; meanwhile, Chloe considers that '*Criollo*' is a good variety for making chocolate bars because of its flavor profile, but it is not the best due to it is rather expensive. Since they are not trained chocolate makers, these friends have focused on different attributes of '*Criollo*' according to what each of them considers to be relevant for deciding whether or not this variety is the best for making chocolate bars. As such, this last situation constitutes an example of a *flexible attribute-set group decision-making* (FAST-GDM) problem [15].

To deal with such group decision-making problems, the participants (experts or nonexperts) can be involved in a *consensus reaching process* (CRP) where they iteratively try to reach a collective agreement on the best option(s) [6, 8, 10]. For example, in the previous case a collective decision can be made using a process where Alice, Bob and Chloe iteratively reconsider their evaluations based on the attributes of the varieties that were initially unobserved by some of them but observed by others. Yet, quantifying the level of concordance between the individual evaluations given by each participant and the collective evaluations computed for the group is regarded as a challenging task.

Aiming to address that challenge, in [16] we conducted a pilot test to get insights about how well theoretical indices proposed to quantify the level of concordance in FAST-GDM problems could reflect the perceived levels. In that work, *intuitionistic fuzzy set contrasting charts*, IFSCC for short, were introduced to facilitate the comparison between the evaluations given by a person and the evaluations computed for the group and, so, make it easy the judgment of the level of concordance. However, the use of IFSCCs that include additional information about the potential decisions was suggested to deal with a case in which, although individual and collective evaluations lead to complete opposite decisions, the perceived level of concordance between them is not the lowest. Following that suggestion, in this paper we propose a novel variant of an IFSCC in which an explicit ranking of the evaluated options is incorporated. We use this variant to perform the aforementioned pilot test with a different group of persons.

To describe the variant of the IFSCCs and present the results of the test, the paper is structured as follows. Preliminary concepts and the formulation of a FAST-GDM problem are presented in the next section. The IFSCCs and the novel variant are described in Sect. 3. The pilot test and the composition of new group of participants are presented in Sect. 4. The results and our findings are presented in Sect. 5. Related work is presented in Sect. 6. The paper in concluded in Sect. 7.

2 Preliminaries

In this section, we first explain the concepts of *augmented appraisal degrees* (AADs) and *augmented (Atanassov) intuitionistic fuzzy sets* (AAIFSs), which have been proposed in [13] to handle experience-based evaluations. Then, we use these concepts in the formulation of the *flexible attribute-set group decision-making* (FAST-GDM) problem introduced in [15]. After that, we present some of the theoretical indices that have been proposed to quantify the level of concordance in FAST-GDM problems.

2.1 Augmented Appraisal Degrees

Let $X = \{x_1, \cdots, x_n\}$ be a collection of potential solutions (called *options*) for a given problem, where each $x_i \in X$ has a collection of features $\mathcal{F}_i$. Let A be a collection of *suitable options* for this problem in X, i.e., $A \subseteq X$. Let p be a proposition having the canonical form 'x_i IS A' meaning 'x_i is a member of A.' Finally, let P be a participant who has been asked to evaluate p. In this context, the evaluation of an option, say x_i, performed by P can be characterized by an *augmented appraisal degree* (AAD) of x_i, say $\hat{\mu}_{A@P}(x_i)$, which is a pair $\langle \mu_{A@P}(x_i), F_{\mu_{A@P}}(x_i) \rangle$ that indicates the level $\mu_{A@P}(x_i)$ to which x_i satisfies p, as well as the collection of features $F_{\mu_{A@P}}(x_i) \in \mathcal{F}_i$ considered by P while appraising p.

By way of illustration, let $X = \{'Criollo', 'Forastero', 'Trinitario'\}$ be a collection of (varieties of) cacao beans that can be used for making chocolate bars; and let A be the collection of *the best varieties for making chocolate bars* that one can be found in X. Assume that a unit interval scale where 0 and 1 represent the lowest and the highest level of satisfiability respectively. With these considerations, the evaluation of (the variety) *'Criollo'* given by Alice in the second introductory example can be characterized by the AAD $\hat{\mu}_{A@Alice}('Criollo') = \langle 0.9, \{'delicate'\}\rangle$.

2.2 Augmented (Atanassov) Intuitionistic Fuzzy Sets

During an evaluation process, a participant can indicate not only *how acceptable* but also *how unacceptable* an option could be – e.g., in the second introductory example, Chloe considered that *'Criollo'* cacao is a good variety for making cacao bars because of its flavor profile, but it is not the best due to its rather expensive. To handle such evaluations, one can characterize them as an *augmented (Atanassov) intuitionistic fuzzy set* (AAIFS) [13], in which the idea of an AAD is integrated with the idea of an *intuitionistic fuzzy set* (IFS) [1, 2] as follows.

Let $X = \{x_1, \cdots, x_n\}$ be a collection of potential solutions for a given problem and let A be a collection of *suitable options* for this problem in X. Consider that each $x_i \in X$ has a collection of features $\mathcal{F}_i$ and assume that $\mathcal{F} = \mathcal{F}_1 \cup \cdots \cup \mathcal{F}_n$. Assume also $\mathcal{I} = [0, 1]$. Let $\hat{\mu}_{A@P}(x_i) = \langle \mu_{A@P}(x_i), F_{\mu_{A@P}}(x_i)\rangle$ and $\hat{v}_{A@P}(x_i) = \langle v_{A@P}(x_i), F_{v_A@P}(x_i)\rangle$ in $\langle \mathcal{I}, \mathcal{F}\rangle$ be two AADs denoting the evaluations given by a person P on *how acceptable* and *how unacceptable* x_i is for fulfilling the proposition 'x_i IS A' respectively. With these considerations, an AAIFS, say $\hat{A}_{@P}$, can be defined by the expression

$$\hat{A}_{@P} = \{\langle x_i, \hat{\mu}_{A@P}(x_i), \hat{v}_{A@P}(x_i)\rangle \mid (x_i \in X)$$
$$\wedge\, (0 \le \mu_{A@P}(x_i) + v_{A@P}(x_i) \le 1)\}. \tag{1}$$

As can be noticed, the *consistency condition*, i.e., $0 \le \mu_{A@P}(x_i) + v_{A@P}(x_i) \le 1$, introduced in the original definition of an IFS has been included in the definition of an AAIFS. Because of this, AAIFSs are deemed to be suitable for the characterization of evaluations that might be marked by hesitation [13].

2.3 FAST-GDM Problem

Since an AAIFS can be used for the characterization of evaluations that might be marked by hesitation, it can be suitable for handling evaluations in a decision-making process in which no constraint on the attributes of the potential solutions has been established. For this reason, the AAIFS concept has been incorporated into the formulation of a FAST-GDM problem as follows [15]:

Let $X = \{x_1, \cdots, x_n\}$ be a collection of *potential options* for a particular problem, and let $A \subseteq X$ be a collection of *suitable options* for this problem. Let $E = \{E_1, \cdots, E_m\}$ be a collection of participants, experts or non-experts, who were asked to evaluate to which level each option in X is member of A. Let

$$\hat{A}_{@E_j} = \{\langle x_i, \hat{\mu}_{A@E_j}(x_i), \hat{v}_{A@E_j}(x_i)\rangle \mid (x_i \in X)$$
$$\wedge \left(0 \leq \mu_{A@E_j}(x_i) + v_{A@E_j}(x_i) \leq 1\right)\} \tag{2}$$

be an AAIFS characterizing the *individual* evaluations given by any $E_j \in E$. Let

$$\hat{A} = \{\langle x_i, \hat{\mu}_A(x_i), \hat{v}_A(x_i)\rangle \mid (x_i \in X)$$
$$\wedge (0 \leq \mu_A(x_i) + v_A(x_i) \leq 1)\} \tag{3}$$

be an AAIFS characterizing the *collective* evaluations computed for (the group of participants) E. Finally, let $\text{cix}(\hat{A}_{@E_j}, \hat{A})$ be a monotonically increasing function, called *concordance index*, that computes the level of concordance between $\hat{A}_{@E_j}$ and $\hat{A}$. In this context, a FAST-GDM problem consists in finding the most suitable option(s) in such a way that the average of all the *concordance indices*, i.e., $\frac{1}{m}\sum_{E_j \in E} \text{cix}(\hat{A}_{@E_j}, \hat{A})$, is maximized.

2.4 Theoretical Concordance Indices

As can be noticed in the above formulation, a crucial part of a FAST-GDM problem is the computation of the level of concordance between individual and collective evaluations. Hence, choosing a reliable method for its computation constitutes an essential task.

In [16], we suggested the computation of the level of concordance by means of a function S that computes the similarity between $\hat{A}_{@E_j}$ and $\hat{A}$. That is to say, we suggested the computation of a concordance index between $\hat{A}_{@E_j}$ and $\hat{A}$ by means of

$$\text{cix}(\hat{A}_{@E_j}, \hat{A}) = S(\hat{A}_{@E_j}, \hat{A}). \tag{4}$$

In this regard, the following similarity measures, which are designed to compare traditional IFSs, have been considered for the computation of theoretical concordance indices—the interest reader is referred to [14] for an open-source implementation of these similarity measures:

- XVBr-α [12], which is defined by the equation

$$S_{@A}^{\alpha}(J, A) = \Delta_{@A} \cdot S^{\alpha}(J, A), \tag{5}$$

where $\Delta_{@A} \in [0, 1]$ is a factor that can be computed through the method *spotRatios* proposed in [12] and $S^{\alpha}(J, A)$ is given by

$$S^{\alpha}(J, A) = 1 - \frac{1}{n} \sum_{i=1}^{n} \left| (\mu_A(x_i) - \mu_J(x_i)) \right. \tag{6}$$
$$\left. + \alpha (h_A(x_i) - h_J(x_i)) \right| ;$$

– *SK1* [19], which is given by

$$S_{SK1}(J, A) = 1 - f\left(l(J, A), l(J, A^c) \right); \tag{7}$$

– *SK2* [19], which is given by

$$S_{SK2}(J, A) = \frac{1 - f\left(l(J, A), l(J, A^c) \right)}{1 + f\left(l(J, A), l(J, A^c) \right)}; \tag{8}$$

– *SK3* [19], which is given by

$$S_{SK3}(J, A) = \frac{(1 - f\left(l(J, A), l(J, A^c) \right))^2}{(1 + f\left(l(J, A), l(J, A^c) \right))^2} \tag{9}$$

and
– *SK4* [19], which is given by

$$S_{SK4}(J, A) = \frac{e^{-f(l(J,A),l(J,A^c))} - e^{-1}}{1 - e^{-1}}. \tag{10}$$

In Eqs. (7), (8), (9) and (10), A^c is the complement of A, i.e.,

$$A^c = \{\langle x_i, \nu_A(x_i), \mu_A(x_i) \rangle | (x_i \in X)$$
$$\wedge (0 \le \mu_A(x_i) + \nu_A(x_i) \le 1)\}, \tag{11}$$

$l(J, A)$ is a function that measures the distance between A and J [19], e.g., $l(J, A)$ can be the Hamming distance defined by

$$l(J, A) = \frac{1}{2n} \sum_{i=1}^{n} (|\mu_A(x_i) - \mu_J(x_i)| + \tag{12}$$
$$+ |\nu_A(x_i) - \nu_J(x_i)| + |h_A(x_i) - h_J(x_i)|),$$

and

$$f\left(l(J, A), l(J, A^c) \right) = \frac{l(J, A)}{l(J, A) + l(J, A^c)}. \tag{13}$$

It is worth mentioning that a *flat operator* [16], $\dashv\!\!\mid \cdot \mid\!\!\vdash$, which turns an AAIFS into an IFS by excluding the collections of features recorded in each AAIFS element, can be used for converting $\hat{A}_{@E_j}$ and $\hat{A}$ into two IFSs, say J and A respectively.

3 Intuitionistic Fuzzy Set Contrasting Charts

As in [16], our aim in this extended version is to study how well a theoretical concordance index proposed to quantify the level of concordance in FAST-GDM problems could reflect what is perceived. In that work, we proposed a visual representation of an IFS, called *IFS contrasting chart* (IFSCC), to facilitate the interpretation of the evaluations given by a person or computed for a group during a FAST-GDM problem. As a sequel, in this work we introduce a novel variant of an IFSCC which includes information about a potential decision on the evaluated options. Both the original and the novel variant of an IFSCC are described below.

3.1 Idea Behind an IFSCC

In [16] we described the idea behind an IFSCC through the following analogy. Consider a discrete collection $X = \{x_1, \cdots, x_n\}$ of potential solutions for a given problem and imagine that a buoy floating on the surface of the sea represents the evaluation of any x_i in X. Imagine also that an air column and ballast can be contained inside a buoy: while the air column represents the level to which x_i is a suitable option, i.e., $\mu_A(x_i)$, the ballast represent the level to which x_i is an unsuitable option, i.e., $\nu_A(x_i)$. Finally, imagine that, due to the *consistency condition* (see Sect. 2.2), the height of a buoy is limited to the unit interval [0, 1]. In this setting, the *buoyancy* of a buoy, say $\rho_A(x_i)$, is given by

$$\rho_A(x_i) = \mu_A(x_i) - \nu_A(x_i) \tag{14}$$

and denotes the level to which x_i is a *suitable option* ($\rho_A(x_i) > 0$) or an *unsuitable option* ($\rho_A(x_i) < 0$). An example of the evaluation of two options, x_1 and x_2, using this analogy is depicted in Fig. 1 [16]. Notice that, while the buoyancy of x_1 is positive, the buoyancy of x_2 is negative. This suggests that x_1 is more suitable than x_2.

3.2 IFSCCs and AAIFSs

Using the above-mentioned analogy, one can represent the appraisal levels in an AAIFS. For instance, Fig. 2 is an IFSCC that represents an AAIFS, say $\hat{A}_{@P}$, which characterizes the evaluations of the options x_1, x_2, x_3 and x_4 given by a partici-

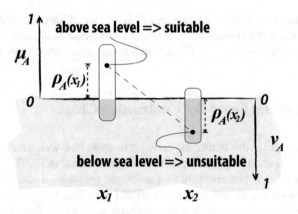

Fig. 1 Idea behind an IFS contrasting chart [16]

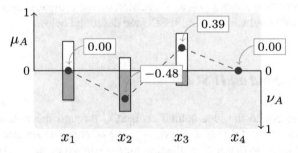

Fig. 2 Evaluations of the options x_1, x_2, x_3 and x_4 given by a participant P

pant P. In this figure, the evaluation of x_1 is represented by a 'buoy' with equal parts of 'air' ($\mu_{A@P}(x_1)$) and 'ballasts' ($\nu_{A@P}(x_1)$) and, thus, its buoyancy is 0, i.e., $\rho_{A@P} = \mu_{A@P}(x_1) - \nu_{A@P}(x_1) = 0$, which suggests that x_1 has some 'positive' features but also some 'negative' features that make P to think that x_1 neither satisfies nor dissatisfies its membership in the collection of suitable options for the problem under evaluation. Analogously, while the 'buoy' related to x_2 has less 'air' than 'ballasts' and, thus, x_2 has a negative buoyancy ($\rho_{A@P}(x_2) = -0.48$), the 'buoy' related to x_3 has more 'air' than 'ballasts' and, thus, x_3 has a positive buoyancy ($\rho_{A@P}(x_3) = 0.39$).

The *hesitation margin* [1, 2], which is given by $h_A(x_i) = 1 - (\mu_A(x_i) + \nu_A(x_i))$, is not explicitly depicted in an IFSCC. However, it can be inferred. For instance, the missing 'buoy' related to x_4 in Fig. 2 suggests that the evaluation of x_4 has been missed or that P did not evaluate x_4. Hence, considering x_4 as a suitable (or unsuitable) option is marked by a high hesitation in this case.

3.3 Including Information About a Potential Decision in IFSCCs

In [16], we found a case in which evaluations lead to complete opposite decisions although the perceived level of concordance between them was not the lowest. The evaluations related to that case are represented by means of the IFSCCs depicted in Fig. 3: while the collective evaluations of the options x_1, x_2 and x_3 are depicted in Fig. 3a, the evaluations given by a participant P are depicted in Fig. 3b. The case is as follows.

To decide which of the evaluated options are the most suitable ones, we can sort the options in descending order according to their buoyancy. After doing so with the evaluations in Fig. 3a, b, we obtain the ordered lists "x_3, x_2, x_1" and "x_1, x_2, x_3" respectively. As can be noticed, these lists lead to opposite decisions. Even so, the average of the perceived level of concordance related to these evaluations is appreciably greater than the expected theoretical value for this case [12].

A potential explanation given in [16] for such a case is that a more clear indication of the potential decision might be needed. For this reason, the incorporation of an explicit ranking of the options in an IFSCC is proposed as is shown in Fig. 4. The rationale behind the inclusion is that such a ranking can help a person to estimate in a better way the level of concordance between the collective and individual evaluations.

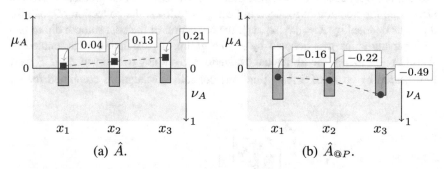

Fig. 3 Evaluations of the options x_1, x_2 and x_3 represented by original IFSCCs: while the collective evaluations are depicted in (**a**), the evaluations given by a participant P are depicted in (**b**)

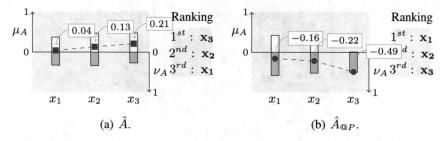

Fig. 4 Evaluations of the options x_1, x_2 and x_3 represented by the proposed variant of IFSCCs

In the next section, we describe how this variant of IFSCC has been used to replicate the pilot test carried out in [16].

4   Pilot Test

In the pilot test described in [16], one can identify two parts, each with a different group of participants: while the first part is about unfolding a consensus reaching process (CRP) in a particular FAST-GDM problem, the second part is concerned with getting the level of concordance that is perceived between the collective and the individual evaluations performed in the first part and, also, with the comparison of those perceived levels with the theoretical concordance indices. We summarize these parts below.

4.1   A Consensus Reaching Process in a FAST-GDM Problem

The FAST-GDM problem presented in [16] was about finding the best smooth dip(s), among 3 potential dips, to pair with banana chips. A group of 11 people, 6 women and 5 men, were asked to solve that problem. The CRP in that problem was modeled using the notation introduced in Sect. 2 as follows: the collection of potential dips was denoted by $X = \{x_1, x_2, x_3\}$; the collection of the best smooth dip to pair with banana chips was denoted by A; the group of participants was denoted by $E = \{E_1, \cdots, E_{11}\}$; the individual evaluations given any participant $E_j \in E$ were represented by an AAIFS $\hat{A}_{@E_j}$; and the collective evaluations computed for the

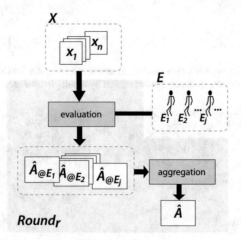

Fig. 5 Getting augmented evaluations [16]

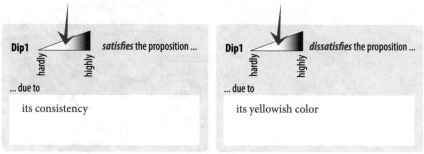

Fig. 6 Form for recording an augmented evaluation

group were represented by an AAIFS $\hat{A}$ (see Fig. 5 [16]). Two rounds were completed by these 11 persons. This means that 12 AAIFSs (1 AAIFS representing the collective evaluations and 11 AAIFSs representing the individual evaluations) were obtained during that CRP.

It is worth mentioning that those persons were provided with a form like the one shown in Fig. 6 to get such augmented evaluations. Notice that, a participant can indicate how suitable and how unsuitable an option could be, as well as (some of) the reasons that justify his/her appraisal in such a form.

The evaluations of the potential options filled out by a person, say P, were characterized as an AAIFS by means of the following procedure. Consider that the levels of suitability and unsuitability filled out by P for any $x_i \in X$ can be linked to two values in a unit interval scale, say $\check{\mu}_{A@P}(x_i)$ and $\check{\nu}_{A@P}(x_i)$ respectively, where 1 denotes the highest value and 0 the lowest. Also, consider that the features filled out for any $x_i \in X$ can be included in two collections, say $F_{\mu_{A@P}}(x_i)$ and $F_{\nu_{A@P}}(x_i)$. With these considerations, one can obtain an AAIFS element $\langle x_i, \hat{\mu}_{A@P}(x_i), \hat{\nu}_{A@P}(x_i) \rangle$ for each option $x_i \in X$ such that

$$\hat{\mu}_{A@P}(x_i) = \langle \check{\mu}_{A@P}(x_i)/\eta, \ F_{\mu_{A@P}}(x_i) \rangle, \tag{15}$$

$$\hat{\nu}_{A@P}(x_i) = \langle \check{\nu}_{A@P}(x_i)/\eta, \ F_{\nu_{A@P}}(x_i) \rangle \tag{16}$$

and

$$\eta = \max(1, (\check{\mu}_{A@P}(x_i) + \check{\nu}_{A@P}(x_i))). \tag{17}$$

Regarding the quantification of the concordance indices, they were computed by means of (4), where S was chosen among five similarity measures presented in Sect. 2.4 as depicted in Fig. 7 [16].

Fig. 7 Quantification of theoretical concordance indices [16]

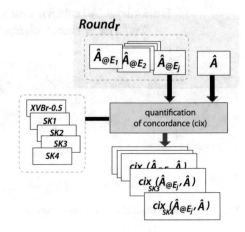

4.2 Getting and Comparing Perceived Levels of Concordance

To obtain the perceived level of concordance between the individual and the collective evaluations in the above-mentioned CRP, we first built the variants of IFSCCs (see Sect. 3.3) corresponding to the 24 AAIFSs that characterize those evaluations (see those IFSCCs in Figs. 13 and 14). Then, we made 22 pairs of IFSCCs in such a way that the individual evaluations given by each participant could be compared with the computed collective evaluations in each round.

Fig. 8 Quantification of perceived levels of concordance [16]

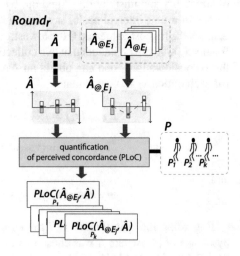

Using those 22 pairs of IFSCCs, in this extended work we asked a group of 49 persons pursuing a Master's degree in Information Management to indicate the level of concordance that each of them perceives between each pair of IFSCCs. This process is depicted in Fig. 8 [16].

To compare the perceived levels of concordance given by the group of 49 people with the theoretical concordance indices, we followed both a *macro* approach and a *micro* approach: while in the macro approach the average and the median of the perceived levels of concordance are individually contrasted with each concordance index, in the micro approach each perceived level of concordance is contrasted with each concordance index. The procedures for computing those values are described below.

Let $P = \{P_1, \cdots, P_{49}\}$ be a collection that represents the group of the 49 persons and consider that $\text{PLoC}_{P_k}(\hat{A}_{@E_j}, \hat{A})$ represents the level of concordance between $\hat{A}_{@E_j}$ and $\hat{A}$ perceived by any $P_k \in P$.

In the macro approach, the average of the perceived levels for each pair $\langle \hat{A}_{@E_j}, \hat{A} \rangle$ is computed by

$$\overline{\text{PLoC}}(\hat{A}_{@E_j}, \hat{A})) = \frac{1}{|P|} \sum_{P_k \in P} (\text{PLoC}_{P_k}(\hat{A}_{@E_j}, \hat{A})) \tag{18}$$

while the median is given by

$$\widetilde{\text{PLoC}}(\hat{A}_{@E_j}, \hat{A})) = \text{med}\{\text{PLoC}_{P_k}(\hat{A}_{@E_j}, \hat{A})\}, \forall P_k \in P. \tag{19}$$

The absolute error between the average of the perceived levels of concordance and a given concordance index, say cix, is given by

$$\Delta_{\text{cix}}(\hat{A}_{@E_j}, \hat{A}) = |\overline{\text{PLoC}}(\hat{A}_{@E_j}, \hat{A}) - \text{cix}(\hat{A}_{@E_j}, \hat{A})|, \tag{20}$$

where $|\cdot|$ denotes the absolute value. Likewise, the absolute error between the median of the perceived levels of concordance and cix is given by

$$\ddot{\Delta}_{\text{cix}}(\hat{A}_{@E_j}, \hat{A}) = |\widetilde{\text{PLoC}}(\hat{A}_{@E_j}, \hat{A}) - \text{cix}(\hat{A}_{@E_j}, \hat{A})|. \tag{21}$$

A *macro mean absolute error* is computed by

$$\overline{\Delta}_{\text{cix}} = \frac{1}{|E|} \sum_{E_j \in E} \Delta_{\text{cix}}(\hat{A}_{@E_j}, \hat{A}), \tag{22}$$

where $|E|$ represents the number of participants in E. Analogously, a *macro median absolute error* is computed by

$$\widetilde{\Delta}_{\text{cix}} = \text{med}\{\ddot{\Delta}_{\text{cix}}(\hat{A}_{@E_j}, \hat{A})\}, \forall E_j \in E. \tag{23}$$

In both cases, values of $\overline{\Delta}_{\text{cix}}$ and $\widetilde{\Delta}_{\text{cix}}$ close to 0 means that cix reflects well the perceived level of concordance.

Regarding the micro approach, the absolute error between the perceived level of concordance and a given concordance index, say cix, is computed for each pair $\langle \hat{A}_{@E_j}, \hat{A} \rangle$ and for each person $P_k \in P$, by means of

$$\delta_{\text{cix}, P_k}(\hat{A}_{@E_j}, \hat{A}) = | \text{PLoC}_{P_k}(\hat{A}_{@E_j}, \hat{A})) - \text{cix}(\hat{A}_{@E_j}, \hat{A})|. \tag{24}$$

While a *micro mean absolute error* is given by

$$\overline{\delta}_{\text{cix}} = \frac{1}{|E| \times |P|} \sum_{E_j \in E, P_k \in P} \delta_{\text{cix}, P_k}(\hat{A}_{@E_j}, \hat{A}), \tag{25}$$

a *micro median absolute error* is given by

$$\widetilde{\delta}_{\text{cix}} = \text{med}\{\delta_{\text{cix}, P_k}(\hat{A}_{@E_j}, \hat{A})\}, (\forall E_j \in E) \wedge (\forall P_k \in P). \tag{26}$$

Analogously to $\overline{\Delta}_{\text{cix}}$ and $\widetilde{\Delta}_{\text{cix}}$, values of $\overline{\delta}_{\text{cix}}$ and $\widetilde{\delta}_{\text{cix}}$ close to 0 means that cix reflects well the perceived level of concordance.

It is worth mentioning that, while we introduced the equations (18), (20), (22), (24) and (25) in [16], in this work we introduce the equations (19), (21), (23) and (26). In the next section, we present the results obtained after using these equations.

5   Results and Discussion

5.1   Results

Tables 1 and 2 summarize the results obtained after computing the theoretical concordance indices, identified by *XVBr-0.5*, *SK1*, *SK2*, *SK3* and *SK4*, as well as the macro averages and the macro medians of the perceived levels of concordance, denoted by $\overline{\text{PLoC}}$ and $\widetilde{\text{PLoC}}$ respectively: while Table 1 shows the results related to the evaluations performed during first round (see Fig. 13 in Appendix), Table 2 shows the results related to second round (see Fig. 14 in Appendix).

The data in those tables were used as input of (20) and (21) to compute the absolute errors between the average of the perceived levels of concordance (PLoCs) and each concordance index, as well as the absolute errors between the median of the PLoCs and each concordance index.

Table 1 Theoretical concordance indices along with the average and the median of the perceived levels of concordance (Round 1)

Pair	XVBr-0.5	SK1	SK2	SK3	SK4	$\overline{\text{PLoC}}$	$\widetilde{\text{PLoC}}$
$\langle \hat{A}_{@E_1}, \hat{A} \rangle$	0.75	0.58	0.41	0.17	0.46	0.37	0.30
$\langle \hat{A}_{@E_2}, \hat{A} \rangle$	0.19	0.48	0.32	0.10	0.36	0.19	0.12
$\langle \hat{A}_{@E_3}, \hat{A} \rangle$	0.59	0.41	0.26	0.07	0.29	0.52	0.50
$\langle \hat{A}_{@E_4}, \hat{A} \rangle$	0.00	0.42	0.27	0.07	0.30	0.35	0.30
$\langle \hat{A}_{@E_5}, \hat{A} \rangle$	0.76	0.59	0.42	0.18	0.47	0.42	0.40
$\langle \hat{A}_{@E_6}, \hat{A} \rangle$	0.45	0.55	0.38	0.14	0.43	0.40	0.40
$\langle \hat{A}_{@E_7}, \hat{A} \rangle$	0.88	0.55	0.38	0.14	0.42	0.40	0.40
$\langle \hat{A}_{@E_8}, \hat{A} \rangle$	0.50	0.53	0.36	0.13	0.41	0.33	0.33
$\langle \hat{A}_{@E_9}, \hat{A} \rangle$	0.25	0.47	0.31	0.10	0.35	0.26	0.20
$\langle \hat{A}_{@E_{10}}, \hat{A} \rangle$	0.74	0.54	0.37	0.14	0.42	0.49	0.40
$\langle \hat{A}_{@E_{11}}, \hat{A} \rangle$	0.43	0.52	0.35	0.12	0.40	0.41	0.40

Table 2 Theoretical concordance indices along with the average and the median of the perceived levels of concordance (Round 2)

Pair	XVBr-0.5	SK1	SK2	SK3	SK4	$\overline{\text{PLoC}}$	$\widetilde{\text{PLoC}}$
$\langle \hat{A}_{@E_1}, \hat{A} \rangle$	0.43	0.49	0.33	0.11	0.37	0.39	0.30
$\langle \hat{A}_{@E_2}, \hat{A} \rangle$	0.40	0.52	0.35	0.13	0.40	0.22	0.10
$\langle \hat{A}_{@E_3}, \hat{A} \rangle$	0.00	0.34	0.21	0.04	0.24	0.24	0.20
$\langle \hat{A}_{@E_4}, \hat{A} \rangle$	0.20	0.50	0.34	0.11	0.38	0.17	0.10
$\langle \hat{A}_{@E_5}, \hat{A} \rangle$	0.68	0.61	0.43	0.19	0.48	0.44	0.40
$\langle \hat{A}_{@E_6}, \hat{A} \rangle$	0.55	0.65	0.48	0.23	0.53	0.54	0.60
$\langle \hat{A}_{@E_7}, \hat{A} \rangle$	0.54	0.66	0.49	0.24	0.54	0.58	0.70
$\langle \hat{A}_{@E_8}, \hat{A} \rangle$	0.53	0.45	0.29	0.08	0.33	0.39	0.30
$\langle \hat{A}_{@E_9}, \hat{A} \rangle$	0.50	0.61	0.44	0.19	0.49	0.43	0.40
$\langle \hat{A}_{@E_{10}}, \hat{A} \rangle$	0.53	0.49	0.33	0.11	0.37	0.37	0.30
$\langle \hat{A}_{@E_{11}}, \hat{A} \rangle$	0.56	0.65	0.48	0.23	0.53	0.37	0.30

A statistical description of the distribution of the macro median absolute errors computed for the two rounds is shown in Table 3 and Fig. 9. Notice that the macro median absolute errors computed for $SK2$, i.e., $\widetilde{\Delta}_{SK2} = 0.04$, and for $SK3$, i.e., $\widetilde{\Delta}_{SK3} = 0.21$, suggest that $SK2$ reflects the perceived level of concordance better than $SK3$ does so.

A frequency distribution of the absolute errors corresponding to the two rounds are depicted in Fig. 10. As an example, in Fig. 10c it is shown that, while 15 out of 22 (i.e., 68.18%) of the computed absolute errors between SK2 and $\widetilde{\text{PLoC}}$ are

Table 3 Statistical description of macro median absolute errors

Measure	XVBr-0.5	SK1	SK2	SK3	SK4
Median ($\widetilde{\Delta}_{cix}$)	0.19	0.19	**0.04**	0.21	0.08
Average	0.20	0.20	**0.09**	0.20	0.11
Q1	0.09	0.14	**0.03**	0.12	0.03
Q3	0.30	0.27	**0.19**	0.26	0.17
IQR	0.21	0.13	**0.16**	0.14	0.14
Max	0.48	0.42	**0.25**	0.46	0.30
Min	0.03	0.04	**0.01**	0.01	0.00
Skew	0.55	0.68	**0.83**	0.32	0.77

Fig. 9 Statistical description of macro median absolute errors

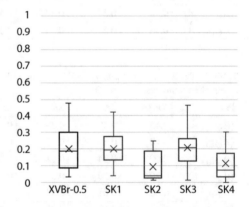

placed in the interval $[0, 0.1]$, 4 out of 22 (i.e., 18.18%) are located in the interval $(0.1, 0.2]$. The values computed by (23) are also shown in Fig. 10. For instance, the macro median absolute error between SK2 and $\widetilde{\text{PLoC}}$, i.e., $\widetilde{\Delta}_{SK2} = 0.04$, is indicated in the center of Fig. 10c.

With respect to the comparisons using a micro approach, the statistical description of micro median absolute errors is shown in Table 4 and Fig. 11. As was done in the macro approach, a frequency distribution of the results computed by (26) are depicted in Fig. 12. In Fig. 12c, e.g., it is shown that, while 133 out of 539 (i.e., 24.67%) of the computed median absolute errors between $SK2$ and PLoC are located in the interval $[0, 0.1]$, 121 out of 539 (i.e., 22.45%) are placed in the interval $(0.1, 0.2]$. In this case, the micro median absolute error between SK2 and PLoC, i.e., $\widetilde{\delta}_{SK2} = 0.21$, is indicated in the center of Fig. 12c. The overall results are summarized in Table 5.

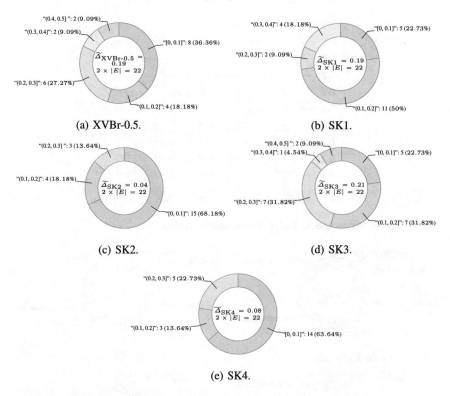

(a) XVBr-0.5.

(b) SK1.

(c) SK2.

(d) SK3.

(e) SK4.

Fig. 10 Distribution of median abs. errors (macro comparisons)

Table 4 Statistical description of micro median absolute errors

Measure	XVBr-0.5	SK1	SK2	SK3	SK4
Median ($\widetilde{\delta}_{\text{cix}}$)	0.23	0.25	0.21	0.21	0.22
Average	0.27	0.26	0.22	0.27	0.22
Q1	0.12	0.14	0.11	0.08	0.11
Q3	0.37	0.38	0.32	0.46	0.33
IQR	0.25	0.24	0.21	0.38	0.22
Min	0.48	0.42	0.25	0.46	0.30
Max	0.88	0.65	0.67	0.89	0.63
Skew	0.81	0.19	0.52	0.77	0.31

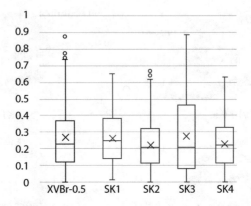

Fig. 11 Statistical description of micro median absolute errors

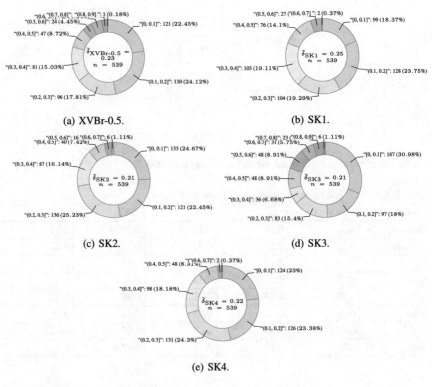

Fig. 12 Distribution of median abs. errors (micro comparisons)

Table 5 Macro and micro mean and median absolute errors

Measure	XVBr-0.5	SK1	SK2	SK3	SK4
$\overline{\Delta}_{cix}$	0.16	0.16	**0.07**	0.24	0.08
$\overline{\delta}_{cix}$	0.27	0.26	**0.22**	0.27	**0.22**
$\widetilde{\Delta}_{cix}$	0.19	0.19	**0.04**	0.21	0.08
$\widetilde{\delta}_{cix}$	0.23	0.25	**0.21**	**0.21**	0.22

5.2 Discussion

In a broad sense, the results listed in Table 5 are consistent with the results obtained in [16]. However, as far as one can see, *SK2* moderately outperforms *XVBr-0.5*, *SK1* and *SK4* according to both the macro mean and the macro median absolute errors. As can be noticed, while the expression $(\overline{\Delta}_{SK2} = 0.07) < (\overline{\Delta}_{SK4} = 0.08) < (\overline{\Delta}_{SK1} = 0.16) \le (\overline{\Delta}_{XVBr-0.5} = 0.16)$ holds for the macro mean absolute errors, $(\widetilde{\Delta}_{SK2} = 0.04) < (\widetilde{\Delta}_{SK4} = 0.08) < (\widetilde{\Delta}_{SK1} = 0.19) \le (\widetilde{\Delta}_{XVBr-0.5} = 0.19)$ holds for the macro median absolute errors. Therefore, we can say that *SK2* has a moderate advantage when reflecting the perceived levels of concordance between the individual and the collective evaluations. In this regard, a practical implication is that *SK2* can be accepted as good indicator of the level of concordance in FAST-GDM problems.

As happened in [16] and although the IFSCCs were augmented with explicit rankings of the options, the macro median and the macro mean of the perceived level of concordance related to the pair $\langle \hat{A}_{@E_4}, \hat{A} \rangle$, i.e., $\widetilde{\mathrm{PLoC}}(\hat{A}_{@E_4}, \hat{A}) = 0.30$ and $\overline{\mathrm{PLoC}}(\hat{A}_{@E_4}, \hat{A}) = 0.35$ respectively, are appreciably greater than the expected theoretical value for this case [12], i.e., $\mathrm{cix}(\hat{A}_{@E_4}, \hat{A}) = 0$. Notice that the individual evaluations characterized by $\hat{A}_{@E_4}$ (see Fig. 13e) and the collective evaluations characterized by $\hat{A}$ (see Fig. 13a) can lead to potential opposite decisions about the options x_1, x_2 and x_3. As was mentioned in [16], a more clear indication of the potential decision that can be taken after studying the evaluations represented in an IFSCC might be still needed. However, it is worth mentioning that the representation of evaluations characterized as IFSs was introduced to the group of participants just a few minutes before obtaining their answers. Therefore, conducting another study in which the perceived levels of concordance would be given by a group of IFSCC-trained participants is suggested.

6 Related Work

In the literature, one can find several geometrical interpretations of IFSs. One of them is the *standard interpretation*, in which the membership and the nonmembership components, i.e., $\mu_A(x)$ and $\nu_A(x)$, are depicted in a shared column of height 1.

A modification of this interpretation, called *unit segment representation*, represents $(1 - v_A(x))$ instead of $v_A(x)$. Another representation, called *IFS-interpretational triangle*, is based on a right triangle having two sides of length 1 corresponding respectively to the membership and nonmembership components [1, 2]. As an advantage, an operation over an IFS element can, in general, be easily visualized in an IFS-interpretational triangle. Nevertheless, a holistic view of all the IFS elements can be unclear in that representation. A variant of the IFS-interpretational triangle has been introduced in [19]. In that variant, the idea of a *'unit cube'* is proposed in order to represent the hesitation margin . A business-oriented representation of an IFS, in which an IFS is depicted like a radar chart, has been proposed in [3]. A potential drawback with that representation is that the membership and nonmembership components are depicted in the same circle (or band), which makes difficult the depiction of the buoyancy. As an alternative to that representation, we proposed in [16] a novel business-oriented representation of an IFS by means of an IFSCC. In an IFSCC, two bands are used (one for each component) to depict in a holistic way the buoyancy of each element in an IFS.

With respect to the quantification of the level of concordance, one can find in the literature methods by which the level of concordance (or agreement) between the evaluations given by two individuals is computed by means of a similarity (or distance) measure defined in the IFS framework. As an example, Szmidt and Kacprzyk have proposed the use of similarity (or distance) measures between two IFSs to compute the level of agreement between two participants whose evaluations have been characterized as IFSs [17–19]. However, to the best of our knowledge, the work presented in [16] and this work are the pioneers in performing empirical studies oriented to determine how well the results computed by such similarity measures reflect the perceived levels of concordance in a consensus reaching process.

7   Conclusions

In this paper, we have introduced a novel variant of the visual representation of an *intuitionistic fuzzy set* (IFS) [1, 2], called *IFS contrasting chart* (IFSCC) [16], which has been proposed to facilitate an estimation of the level of concordance between individual and collective evaluations characterized respectively by two IFSs. In this variant, the original version of an IFSCC has been augmented with the ranking of the evaluated options in order to facilitate the estimation of the level of concordance during a consensus reaching process in a *flexible attribute-set group decision-making* (FAST-GDM) problem [15].

Such augmented IFSCCs have been used to replicate the pilot study presented in [16], in which several theoretical concordance indices based on similarity measures designed to compare *intuitionistic fuzzy sets* (IFSs) were tested in order to determine their usability in FAST-GDM problems.

Although in a broad sense the results in this study are consistent with the results obtained in [16], we found that the concordance index based on the similarity measure $SK2$ [19] moderately outperforms the concordance indices based on the similarity measures $XVBr$-0.5 [12], $SK1$ [19] and $SK4$ [19]. However, we also found that the case reported in [16], in which the perceived level of concordance between the individual and collective evaluations is not the lowest even though such evaluations lead to complete opposite decisions, still occurs.

A likely explanation for the aforementioned case is that the participants who estimated the levels of concordance were not trained extensively about how an augmented IFSCC should be interpreted. In this regard, conducting another study in which the perceived levels of concordance would be given by a group of IFSCC-trained participants is suggested.

Other (still) suggested studies concern: (i) the use of IFSCCs that have been augmented with additional information like a linguistic summary about a potential decision [9]; and (ii) the use of scales of measurement formed from linguistic labels such as *'highly concordant'* or *'hardly concordant'* to map or express the results of the concordance indices, as well as the perceived levels of concordance [6, 7].

Appendix

Evaluations About the 'Best Smooth Dip'

This appendix presents the IFS contrasting charts (IFSCCs) of the evaluations given by 11 persons who tried to reach a consensus about the best smooth dip(s), among 3 potential dips, to pair with banana chips. In contrast to the IFSCCs used in [16], these IFSCCs have been augmented with the ranking of the options.

Figures 13 and 14 show the IFSCCs corresponding to the first round and the second round respectively. Figures 13a and 14a represent the collective evaluations computed for the group during the first and the second round respectively.

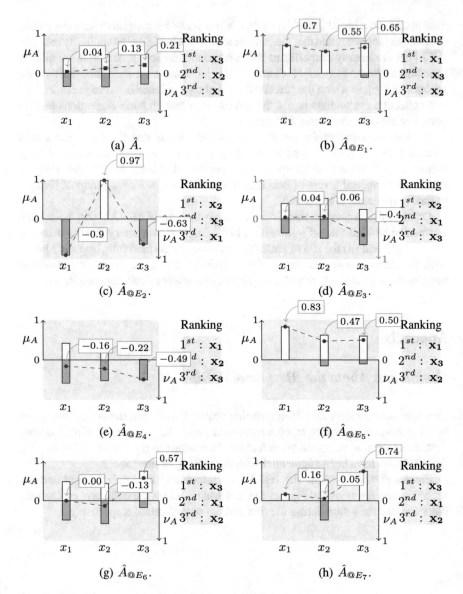

Fig. 13 IFSCCs characterizing the evaluations obtained during Round 1

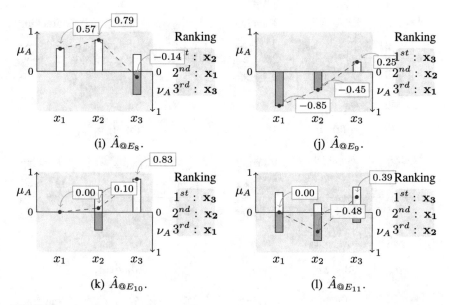

(i) $\hat{A}_{@E_8}$.

(j) $\hat{A}_{@E_9}$.

(k) $\hat{A}_{@E_{10}}$.

(l) $\hat{A}_{@E_{11}}$.

Fig. 13 (continued)

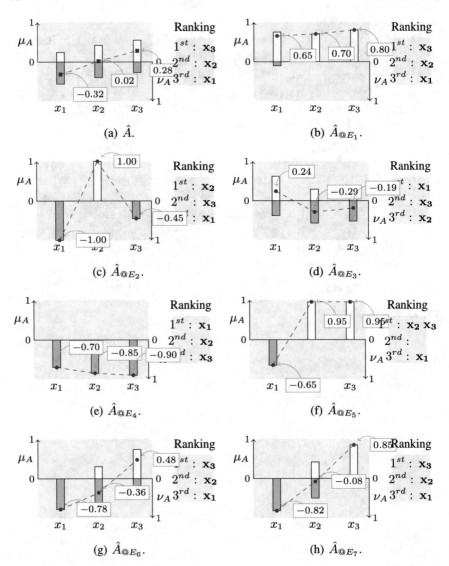

Fig. 14 IFSCCs characterizing the evaluations obtained during Round 2

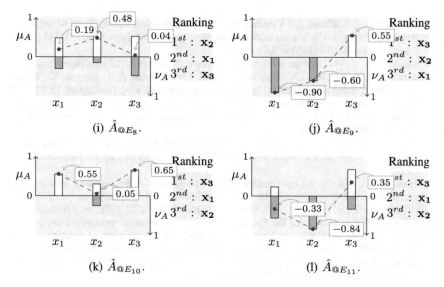

(i) $\hat{A}_{@E_8}$.   (j) $\hat{A}_{@E_9}$.

(k) $\hat{A}_{@E_{10}}$.   (l) $\hat{A}_{@E_{11}}$.

Fig. 14 (continued)

References

1. Atanassov, K.T.: Intuitionistic fuzzy sets. Fuzzy Sets Syst. **20**(1), 87–96 (1986)
2. Atanassov, K.T.: On intuitionistic fuzzy sets theory. In: Studies in Fuzziness and Soft Computing, vol. 283. Springer Berlin Heidelberg, Berlin, Heidelberg (2012). https://doi.org/10.1007/978-3-642-29127-2
3. Atanassova, V.: Representation of fuzzy and intuitionistic fuzzy data by radar charts. Notes Intuit. Fuzzy Sets **16**(1), 21–26 (2010)
4. Bouyssou, D., Dubois, D., Prade, H., Pirlot, M.: Decision Making Process: Concepts and Methods. Wiley (2013)
5. Dong, Y., Xiao, J., Zhang, H., Wang, T.: Managing consensus and weights in iterative multiple-attribute group decision making. Appl. Soft Comput. **48**, 80–90 (2016). https://doi.org/10.1016/j.asoc.2016.06.029
6. Herrera, F., Herrera-Viedma, E., Verdegay, J.L.: A model of consensus in group decision making under linguistic assessments. Fuzzy Sets Syst. **78**(1), 73–87 (1996). https://doi.org/10.1016/0165-0114(95)00107-7
7. Herrera, F., Herrera-Viedma, E., Verdegay, J.L.: A rational consensus model in group decision making using linguistic assessments. Fuzzy Sets Syst. **88**(1), 31–49 (1997). https://doi.org/10.1016/S0165-0114(96)00047-4
8. Kacprzyk, J., Fedrizzi, M.: A 'soft' measure of consensus in the setting of partial (fuzzy) preferences. Eur. J. Oper. Res. **34**(3), 316–325 (1988). https://doi.org/10.1016/0377-2217(88)90152-X
9. Kacprzyk, J., Zadrożny, S.: Linguistic database summaries and their protoforms: towards natural language based knowledge discovery tools. Inf. Sci. **173**(4), 281 – 304 (2005). https://doi.org/10.1016/j.ins.2005.03.002 dealing with Uncertainty in Data Mining and Information Extraction
10. Kacprzyk, J., Zadrożny, S.: Soft computing and web intelligence for supporting consensus reaching. Soft Compu. **14**(8), 833–846 (2010). https://doi.org/10.1007/s00500-009-0475-4

11. Liu, Y., Fan, Z.P., Zhang, X.: A method for large group decision-making based on evaluation information provided by participators from multiple groups. Inf. Fusion **29**, 132–141 (2016). https://doi.org/10.1016/j.inffus.2015.08.002
12. Loor, M., De Tré, G.: In a quest for suitable similarity measures to compare experience-based evaluations. In: Merelo, J.J., Rosa, A., Cadenas, J.M., Correia, A.D., Madani, K., Ruano, A., Filipe, J. (eds.) Computational Intelligence: International Joint Conference, IJCCI 2015 Lisbon, Portugal, November 12–14, 2015, Revised Selected Papers. Studies in Computational Intelligence, vol. 669, pp. 291–314. Springer International Publishing (2017). https://doi.org/10.1007/978-3-319-48506-5
13. Loor, M., De Tré, G.: On the need for augmented appraisal degrees to handle experience-based evaluations. Appl. Soft Comput. **54C**, 284–295 (2017). https://doi.org/10.1016/j.asoc.2017.01.009
14. Loor, M., De Tré, G.: An open-source software package to assess similarity measures that compare intuitionistic fuzzy sets. In: 2017 IEEE International Conference on Fuzzy Systems (FUZZ-IEEE), pp. 1–6 (July 2017). https://doi.org/10.1109/FUZZ-IEEE.2017.8015689
15. Loor, M., Tapia-Rosero, A., De Tré, G.: Refocusing attention on unobserved attributes to reach consensus in decision making problems involving a heterogeneous group of experts. In: Kacprzyk, J., Szmidt, E., Zadrożny, S., Atanassov, K.T., Krawczak, M. (eds.) Advances in Fuzzy Logic and Technology 2017, pp. 405–416. Springer International Publishing, Cham (2018). https://doi.org/10.1007/978-3-319-66824-6
16. Loor, M., Tapia-Rosero, A., De Tré, G.: Usability of concordance indices in FAST-GDM problems. In: Proceedings of the 10th International Joint Conference on Computational Intelligence (IJCCI 2018), pp. 67–78. INSTICC, SciTePress (2018). https://doi.org/10.5220/0006956500670078
17. Szmidt, E., Kacprzyk, J.: Evaluation of agreement in a group of experts via distances between intuitionistic fuzzy preferences. In: Proceedings First International IEEE Symposium Intelligent Systems, vol. 1, pp. 166–170 (2002). https://doi.org/10.1109/IS.2002.1044249
18. Szmidt, E., Kacprzyk, J.: A consensus reaching process under intuitionistic fuzzy preference relations. Int. J. Intell. Syst. **18**(7), 837–852 (2003). https://doi.org/10.1002/int.10119
19. Szmidt, E., Kacprzyk, J.: A concept of similarity for intuitionistic fuzzy sets and its use in group decision making. In: IEEE International Conference on Fuzzy Systems, pp. 1129–1134 (2004). https://doi.org/10.1109/FUZZY.2004.1375570

Artificial Intelligence Algorithms Selection and Tuning for Anomaly Detection

Zoltan Czako, Gheorghe Sebestyen, and Anca Hangan

Abstract Artificial intelligence (AI) algorithms have recently become a topic of interest, among researchers and commercial product developers. A large number of AI algorithms aiming different kind of real-life problems have qualitatively different results, based on the nature of data, the nature of the problems and based on the context in which they are used. Selecting the most appropriate algorithm to solve a particular problem is not an easy task. In our research, we focus on developing methods and instruments for the selection and tuning of AI algorithms, to solve the problem of anomaly detection in datasets. The goal of our research is to create a platform, which can be used for data analysis. With this platform, the user could be able to quickly train, test and evaluate several artificial intelligence algorithms and also they will be able to find out which is the algorithm that performs best for a specific problem. Moreover, this platform will help developers to tune the parameters of the chosen algorithm in order to get better results on their problem, in a shorter amount of time.

1 Introduction

Every second, huge amounts of data are being generated by devices and applications connected to the Internet such as sensor networks, IoT and wearable devices, smart phones, Google, Facebook and many others. Extracting valuable information or knowledge from this data is not a trivial task and in the absence of some intuitive

Z. Czako (✉) · G. Sebestyen · A. Hangan
Technical University of Cluj-Napoca, Cluj-Napoca, Romania
e-mail: zoltan.czako@cs.utcluj.ro

G. Sebestyen
e-mail: gheorghe.sebestyen@cs.utcluj.com

A. Hangan
e-mail: anca.hangan@cs.utcluj.com

C. Sabourin et al. (eds.), *Computational Intelligence*, Studies in Computational Intelligence 893, https://doi.org/10.1007/978-3-030-64731-5_4

processing rules, the artificial intelligence algorithms are the most feasible solutions. But the question is which type of algorithm performs best in a given case, for a given goal and what is its best parameter setup (e.g. number of neurons and layers for a neural network).

Raw data, however, may contain erroneous or abnormal values for a number of reasons (sensor defects, communication errors, malicious attacks, etc.) and considering them may cause issues in further processing tasks. Therefore, a first step in a data processing flow that may require artificial intelligence is detection and elimination of abnormal values (also called outliers) [6]. In other cases, it is required to automatically recognise some abnormal behaviors of a monitored or controlled system. Again, in many cases there are no precise rules that establish a normal or an abnormal behavior. Through a training process, AI algorithms can learn and then recognise data values or system behaviors as normal or abnormal. The complexity of the anomaly detection problem is given by the diversity of the application domain, the diversity of anomaly types and the complexity of data. There are hundreds of very specialized anomaly detection methods [1]. Many of those use AI algorithms that are able to identify anomalies through classification or pattern recognition.

Having to choose the right AI algorithm for a specific dataset is not straightforward. One has to find the best algorithm that matches the application domain, dataset characteristics and anomaly type. This process is very time consuming, since one has to make a large number of implementations and experiments in order to tune the algorithm's parameters and evaluate the results. If the results are not satisfactory, the process has to be repeated with another algorithm.

In this context, we have created a platform that offers a set of tools which can be used to test multiple AI algorithms and evaluate them using appropriate metrics, this way reducing the time needed to go through the process of choosing an algorithm. Moreover, we propose a methodology for the selection of appropriate AI algorithms for a given problem, by taking into consideration the characteristics of the problem and of the dataset. Finally, we propose a mechanism for automatically tuning the parameters of the chosen subset of algorithms to obtain improved evaluation results for a given dataset.

The anomaly detection platform contains support tools that are needed in different stages of the data analysis process, such as preprocessing (filtering, normalization, etc.), data visualization and performance analysis. Preprocessing of the data can be a challenging task, which may require multiple subsequent tasks. Therefore, our platform contains a set of preprocessing algorithms which can be applied by a domain specialist, without specific programming knowledge. The data can be visualized in multiple ways (2D scatter plots, histograms, FFT, etc.), which can help spotting problems within the raw data. To be able to compare algorithm evaluation results, the platform includes several forms of visualization such as confusion matrix, trend diagram and box-and-whisker plot.

Another important aspect is that an anomaly detection method can combine multiple algorithms to improve the quality of the results. In order to combine different algorithms the platform offers the possibility to pipeline the partial results in a reconfigurable manner.

In the following sections we describe our methodology and give guidelines for selecting a subset of AI algorithms for a specific problem and we discuss the possibility of automating this process. Moreover, we present the mechanism we propose for the automatic tuning of parameters for the selected algorithms. Finally we present our platform and the tools which can be used to evaluate different AI techniques with different types of datasets, configuring the most effective pipeline in different contexts. We show results obtained with different algorithms with and without preprocessing, highlighting in this way the effectiveness of the preprocessing step. We tune and evaluate multiple algorithms and compare the results.

Our tool is very useful in the early stage of a project when the feasibility of a given approach must be measured or decided. The tool will give an estimate of the quality level that may be obtain using different AI algorithms. This will reduce the time to production and it will increase the chance for a successful implementation.

The rest of the paper is organized as follows. Section 2 is a description of related work. Section 3 presents the guidelines for selecting AI algorithms for a problem taking into consideration the problem and dataset characteristics. The mechanism for automatic algorithm parameter tuning is described in Sect. 4. The anomaly detection platform is presented in Sect. 5. Experiments can be found in Sect. 6. Section 7 concludes this paper.

2   Related Work

Choosing the best approach for detecting anomalies in a specific dataset is a real problem. Therefore, there are a lot of surveys that try to organize the large amount of anomaly detection methods. Most of these surveys classify anomaly detection methods based on the application domain (e.g. economical data, networking, industry data, etc.) [1–4] or group methods based on a given theoretical approach (statistics,signal processing, artificial intelligence) [5, 6].

The authors in [1] try to give a solution on how to select anomaly detection methods for a specific application domain. They argue that the selection of a specific anomaly detection technique should be based on anomaly type, the nature of data and on the existence of labeled data. Moreover, they identify the most appropriate detection techniques for each application type.

In [6] the authors focus their classification of anomaly detection methods only on the characteristics of the dataset. They identify three main classes of methods, in which there is no previous knowledge about data (unsupervised), in which there are models only for normality (semi-supervised) and in which there are models for both normal and abnormal behavior (supervised). Furthermore, they make a review of the methods used for anomaly detection that belong tho the three classes they have identified.

The authors of [7] make a comprehensive evaluation of the most popular algorithms used for the problem of credit card fraud detection. This paper evaluates the results of AI algorithms such as Bayesian Networks, Neural Networks, SVM, etc. using accuracy, speed and cost as metrics.

A similar approach for algorithm classification is presented in [8], where the authors consider the characteristics of the available training data as classification criteria.

Many other research papers that discuss algorithm selection guidelines that focus on a specific class of algorithms or on a specific problem category [9, 10].

A common interest that can be observed in these scientific papers is selecting appropriate methods for classes of problems or for classes of datasets. However, many of those algorithms have to be tuned to be able to perform better on specific problems. Tuning algorithm parameters is a time consuming process that can be automatized by using dedicated tools. Google Cloud AutoML [11] is such a tool, that allows users to automatically configure a neural network that best fits their data/problem. Auto-WEKA [12] is a similar tool that provides automatic model selection together with parameter optimization. It considers all algorithms implemented in WEKA and their parameters, and through Bayesian optimization, it identifies the best combination of algorithm and parameters for a dataset.

Compared to existing tools, our approach tries to select a subset of algorithm candidates on which to perform parameter tuning. Moreover, we take into consideration the preprocessing step and visualization support.

3   A Methodology for AI Algorithm Selection

Somehow, every algorithm is context specific; if you try to optimize an algorithm, it will work better for a specific problem, but it will have worse results in the general context. This idea was presented in [13] under the name "No Free Lunch Theorems for Optimization" (NFLT). The NFLT are a set of mathematical proofs and general framework that explores the connection between general-purpose algorithms that are considered "black-box" and the problems they solve. This states that any algorithm that searches for an optimal cost or fitness solution is not universally superior to any other algorithm.

In the real world, we need to decide on engineering solutions to build practical models that solve real problems. For certain pattern recognition problems, we tend to find that certain algorithms perform better than others which is a consequence of the algorithm's fitness or cost function. We are generally not going to find off the shelf algorithms that fits perfectly to our data. We have to tune the algorithm to better fit the data. In order to find optimized solutions, we must have a methodology to architect our model.

In our previous work [14] we introduced a decision tree (Fig. 1) for selecting the right set of algorithms based on the problem context and the nature of the data.

The first step in choosing the appropriate AI algorithm is to determine the nature of the problem and establish the goal. Based on the nature of the problem one can choose one of the following classes of algorithms: Classification, Clustering, Prediction or

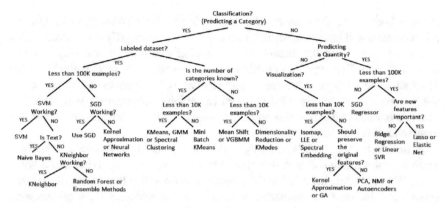

Fig. 1 Algorithm selection decision tree [14]

Visualization (or Dimensionality Reduction). The next step is to focus on the nature of the training data. In this case we should ask questions about the dimension of the available training data, whether the dataset has labels or not or questions about the importance of the features, can we transform the features to get more relevant ones or should we keep the original features. Answering these questions can reduce the search space, in some cases resulting in only one algorithm, this way making the developer's job much easier.

To automatically select a model based on the decision tree presented in Fig. 1 the input dataset must respect a well defined template. The column containing the labels should be the last column and called "outcome". The first row of the dataset must be the header row, and each column should have a name. If the input dataset respects these criteria, an interactive state machine will analyze it by receiving input from the user. Based on the decision tree presented above and by doing some statistical analysis, like calculating the distribution, correlation and other metrics it will match the initial dataset to a subset of the available AI algorithms. User input is necessary in the case of unlabeled data or if the decision can't be inferred based on the dataset.

Along with the selection of AI algorithms, another important factor is the usage of adequate preprocessing methods. There are lots of AI algorithms which require preprocessing of the raw data in order to obtain good results. For example, linear regression or kernel SVM will have better results if the distribution of the attributes from the input data is Gaussian. If the input data contains attributes with skewed distribution, the best way to fix it is to perform a log transform, which will reduce the skewness. Other algorithms have requirements for the density of the input data, the correlation between attributes, variance or even for the magnitude of different attributes. For example, if the dataset contains attributes with different scales, different level of magnitude, then the SVM or the Neural Network will fail or it will have poor result.

Another important characteristic of the input dataset is related to the class distribution. If we have unbalanced data, where the number of examples from one class is much higher than the number of examples from other classes, almost all of the AI algorithms will have poor results, because they will learn only the characteristics of the class for which the dataset contains lots of examples; in this way it won't generalize well. To handle this situation, we can re-sample the input dataset, we can generate some artificial examples or gather more data. There are situations when unbalanced data is exactly what we need (especially for anomaly detection). For example Isolation Forests are based on the assumption that anomalies are few, so the input dataset is unbalanced.

A further problem can be missing data. Usually the input dataset will contain examples with missing data which should be handled before applying an AI algorithm. We can handle the missing data problem by removing those rows (if we have enough training data) or generating artificial data, by replacing missing data with the mean of that column or by replacing it with 0 or other values, depending on the problem. We can also choose algorithms which support missing data, like kNN.

Our platform contains all the necessary preprocessing to handle the situations described above. The preprocessing method selection can be also enhanced by calculating different metrics, like distribution, mean, correlation, variance, feature importance, class distribution and choose the most appropriate preprocessing method as described above.

4 Automatic Tuning of the Selected AI Algorithms

Narrowing down the algorithm selection search space is just the first step of algorithm selection. In the second step we need to find the best parameters (input parameters) for each selected algorithm. This process is very context dependent, for different input data set the algorithms can and will have different behavior and different results. This is why the parameter optimization can be solved only by trying out different values and comparing the results. This could be a very time consuming process and it can have problems with local maximums.

Finding the best parameters for algorithm is basically an optimization problem. This problem is not trivial in case of machine learning algorithms, because each algorithm has different parameters, different behavior and usually the parameters are real numbers, so brute force, exhaustive search is not an option.

In our platform we managed to create a generic solution, which will optimize the parameters of any AI algorithm. In order to solve this optimization problem we used Simulated Annealing. Simulated Annealing can be used to find sub-optimal solution in a discrete search space with a large number of possible solutions (combination of parameters). It is useful for combinatorial optimization problems defined by complex objective functions (model evaluation metrics). This algorithm is a probabilistic technique for approximating global optimum of a given function. In our case the functions are the evaluation metrics. The default evaluation metric is accuracy for classification and mean squared error for prediction , but the user can choose other metrics like, precision, recall, f1-score, mean squared error, r-squared, etc.

As NFLT said, in order to have good results, we must tailor the Simulated Annealing algorithm for our specific problem. We modified the algorithm, and the steps of our algorithm are the following:

1. Select the parameters which will be tuned based on the type of the algorithm (each algorithm has different parameters, see Table 1)
2. Randomly choose multiple values (minimum N > 10 values) for each parameter, and consider each as a separate solution
3. Use Euclidean distance metric to calculate the distance between the randomly selected parameters. If there are parameters for which the distance is less then k (k is the minimum accepted distance between parameters) drop one of the parameters and go to step 2
4. Repeat step 2 and 3 until there are at least N > 10 (N can be increased for a better approximation)
5. Alter the value of each solution by randomly updating value of one parameter by selecting a value in the immediate neighborhood (randomly) to get neighboring state
6. If the combination is already visited, repeat step 5 until a new combination is generated
7. Evaluate model performance on the neighboring state
8. Compare the model performance of neighboring state to the current state and decide whether to accept the neighboring state as current state or reject it based on some criteria (explained below).
9. Based on the results of step 8 repeat step 5 through 8
10. At the end of the algorithm, choose the best solution out of the N selected and tune solution using a proper evaluation metric (by default accuracy for classification and mean squared error for prediction)

Acceptance Criteria:

If the performance of neighboring state is better than current state—Accept

If the performance of neighboring state is worse than current state—Accept with probability $e^{-\beta \Delta f / T}$ where,

β is a constant

T is current temperature

Δ f is the performance difference between the current state and the neighboring state

Note that after a number of **n** iteration the temperature will be modified by **T = T * α**. Because α is less then 1, after each update step the algorithm will accept only better and better solutions, the probability of accepting wrong moves will decrease. If T = 0, the algorithm becomes a greedy algorithm, which accepts only better results than the current result.

Annealing Parameters:

This algorithm takes in four parameters and the effectiveness of the algorithm depends on the choice of these parameters.

Table 1 Parameters of different AI algorithms

Algorithm	Parameters
kNN	n-neighbors
SVM	gamma and C
Random Forest	n-estimators
Isolation Forest	n-estimators
Naive Bayes	var-smoothing
Logistic Regression	C
SGDClassifier	penalty, alpha
AdaBoostClassifier	n-estimators
Neural Network	hidden-layer-size, solver, activation, alpha, learning-rate
KMeans	n-clusters
DBScan	eps, min-samples
AgglomerativeClustering	n-clusters, affinity
AffinityPropagation	damping
PCA	n-components, tol
LDA	n-components, tol
NMF	n-components, tol, alpha

β—normalizing constant Choice of β depends on the expected variation in the performance measure over the search space. In our approach we used $\beta = 1.3$, which is an empiric value.

T_0- initial temperature A good rule of thumb is that your initial temperature T_0 should be set to accept roughly 98

α—Factor by which temperature is scaled after n iterations. This parameter should be greater than 0, but less than 1. Lower values of α restrict the search space at a faster rate than higher values. 0.85 can be chosen by default.

n—number of iterations after which temperature is altered (after every n steps T is updated as T * α. The value of n doesn't affect the results and can be chosen between 5–10. Default value is 5.

maxiters—Number of iterations to perform the parameter search, by deault is 100.

Each algorithm has different parameters, so the first step is to select the proper parameters to be tuned. The parameters of each algorithm can be viewed in Table 1.

These parameters are automatically selected based on the output of the algorithm selection algorithm. Besides the automatic algorithm selection and automatic parameter tuning, our platform permits for the user to manually choose algorithms and manually tune the selected algorithm, this way offering flexible solutions.

5 A Platform for Data Analysis and Anomaly Detection

The Anomaly Detection Platform (ADP) developed by our team offers a pragmatic tool for specialists involved in some kind of data analysis process, with focus on anomaly detection. This tool is meant not as a final solution but as a starting point in the process of finding the best method that fits the quality and efficiency criteria of a given application domain. The platform contains those basic functionalities that allows a specialist to collect, process and visualize data for anomaly detection purposes.

Here are the functionalities we considered necessary for such a platform:

1. **Data Harvesting Tools**: that allow acquisition of data from a variety of datasets having different formats (Excel, CSV, ARFF) or from different physical sources (sensor networks, smart devices, IoT);
2. **Data Preprocessing Tools**: instruments meant to transform the raw data into a noise free and normalized data; typical methods included in this category are: parameterized filters, transforms (e.g. FFT, wavelet) or histograms; there are also methods for determine some statistical parameters of the input data such as: min-max, median value, standard deviation, etc. The preprocessing step is required by some AI algorithms as well as for data visualization;
3. **Artificial Intelligence Algorithms**: contains the most popular AI algorithms. Using this functionality, the user can choose, test and evaluate different types of algorithms manually;
4. **Automatic Algorithm Selection and Tuning Mechanisms**: to find the best algorithm for the given input data, firstly the platform makes some statistical analysis, then applies the simulated annealing algorithm for each selected algorithm to find the best parameters for them. The last step is to evaluate the selected algorithms using the best parameters found and choose one algorithm with the highest score. The algorithms for which the simulated annealing will run can be selected also manually by the user;
5. **Visualization Tools**: very important in the process of finding anomalies in data. They offer a bi-dimensional representation of otherwise multidimensional data, which is much easier to understand for the human eye.We have implemented multiple types of visualization: histogram, time series, scatter plots, line charts; for result comparison: box and whisker plot; for statistical analysis, algorithm comparison and evaluation: correlation matrix, confusion matrix, etc.
6. **Pre-loaded Test Data and Data Generators**: the platform contains pre-loaded benchmark data, which can be used for algorithm evaluation and testing. Special techniques are applied in order to combine "normal" data with artificially created anomalies.

The tools mentioned above are integrated into an open architecture platform allowing continuous extension with new methods. A unique internal predefined data format assures the interoperability and interchangeability between the existing tools and functionalities.

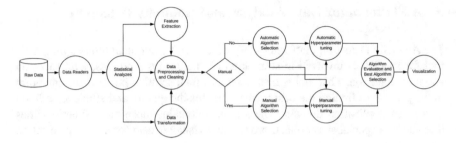

Fig. 2 Pipeline of the algorithm selection and tuning process

We consider that an anomaly detection scenario can be built (in an interactive way) as a pipeline of steps, where the output of a step becomes the input for the next step. The user, through the platform's interface, can combine multiple steps in a single scenario. The pipelines can be saved for later usage.

Figure 2 shows the pipeline architecture of the platform. Each step can be reconfigured, repeated or skipped, the figure shows the default pipeline, which is used in automatic algorithm selection and parameter tuning.

Our platform can be used to visualize the data and the results. We can use scatter plot to view the attributes, line-chart to visualize different trends in the data or in the results, histogram to get a better understanding about the distribution of the attributes, correlation matrix to see attribute-to-attribute or attribute-to-target correlation, confusion matrix to understand the results, etc. as we can see some examples in Fig. 3.

6 Experiments

6.1 Algorithm Selection and Automatic Parameter Tuning

In this experiment we want to show how does our automatic algorithm the selection and parameter tuning work. For this we will use a state of art dataset called Breast Cancer. This is a binary classification, anomaly detection problem, because the output will be yes or no, meaning that the patient has breast cancer or not. The dataset has 569 examples, with 32 attributes and a column which contains the labels.

In the first step, the platform uses the decision tree from Fig. 1 and makes some statistical analysis on the input data, like analyzing the dimensionality of the data set, data types in the data set, feature count to see if the data set is balanced or unbalanced, mean, standard deviation, attribute correlation, minimum and maximum values, 25th, 50th, 75th percentile, class distribution, linearity or non-linearity, importance of features, variance, density, magnitude of different attributes, etc. After this statistical

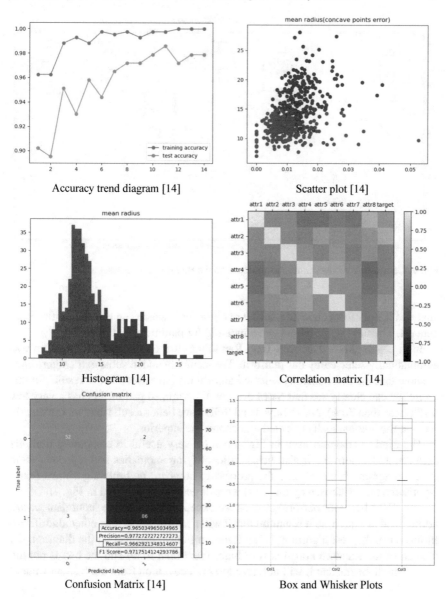

Fig. 3 Visualization examples

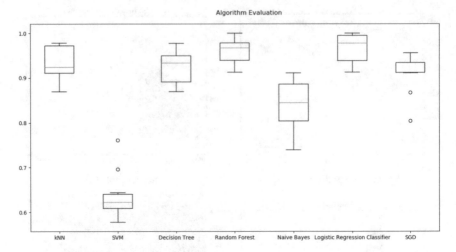

Fig. 4 Comparison of algorithms performances on the Breast Cancer dataset

analyses the platform chooses a subset of algorithms which can be used for the input data set. The next step is simulated annealing for parameter optimization.

In Fig. 4 the kNN, Decision Tree, Random Forest and Logistic Regression was automatically selected by our platform. We manually added some extra algorithms, to show that the automatically selected algorithms have much better results. In this figure we can clearly see that SVM even with optimized parameters has a very low result (less than 70%). Naive Bayes and SGD have better results, but we can clearly see that the automatically selected algorithms are superior.

The kind of diagram used in Fig. 4 can be very useful in comparing multiple algorithms, furthermore it also helps us identifying anomalies and represent them on the diagram. After running the process to compare multiple algorithms (extra algorithms can be chosen by the user) we will get results depicted in Fig. 4.

The SVM has poor performance, as we expected, because the input data wasn't normalized or scaled, so it is natural that it will perform worse than other algorithms. In this case, the best algorithm is the Random Forest, because on the diagram, the median of the Random Forest is very high (more than 95%) and the box is not too large, which means that it will not have too high deviation from the median value.

6.2 Parameter Tuning using Preprocessed Data

The next experiments were made on the same data set as in the previous sub-section, to show how for some algorithms, the preprocessing step can improve their performance. For this, we will use SVM because in the previous experiment this algorithm had inferior results (below 70%). In the case of the SVM, we can configure the C

Table 2 Effect of preprocessing on SVM algorithm [14]

Use Case	Accuracy	Precision	Recall	F1-Score
SVM without preprocessing	0.62	0.62	0.5	0.55
SVM using Min-Max Scaler	0.93	0.97	0.96	0.97

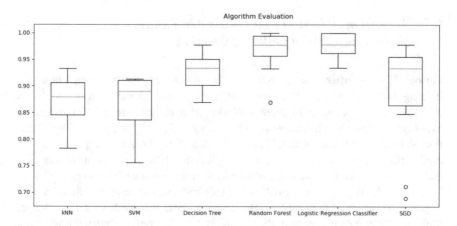

Fig. 5 Comparison of algorithms performances on the Breast Cancer data set using Min-Max Scaler as a preprocessing step

parameter, which stands for penalty parameter of the error term, the value of gamma, which is the kernel coefficient for rbf, poly and sigmoid and the kernel type to be used in the algorithm. We will use the same parameters that were automatically selected by the simulated annealing algorithm. The parameters are C = 10 and gamma = 1. Firstly we ran the SVM using the raw data, without preprocessing and after that we used the Min-Max Scaler in the preprocessing step to scale the raw data. The results are shown in Table 2.

After this experiment we can conclude that, in this case, the preprocessing step makes a big difference on the evaluation results. However, the same preprocessing step does not fit with other algorithms.

In Fig. 5 we can see the generated box and whisker plots after the preprocessing step was applied. From this figure we can see that the performance of the SVM was highly increased, the accuracy is over 90

If we compare Fig. 5 with Fig. 4 we can see that the accuracy of the kNN has dropped, so the preprocessing step decreased its results.

When choosing a preprocessing algorithm we should have in mind the AI algorithm which will be used, the magnitude difference of the attributes, the distribution of the the attributes and other statistical mertrics as we described in Sect. 3.

Table 3 Performance comparison for SVM tuning

SVM Parameters	TP	TN	FP	FN	Accuracy	Precision	Recall	F1-Score
C = 1, Gamma = 0.1	918	0	23	2	0.976	1.0	0.08	0.148
C = 10, Gamma = 1	916	2	8	17	0.989	0.895	0.68	0.773
C = 100, Gamma = 10	915	3	5	20	0.992	0.87	0.8	0.833

6.3   Algorithm Selection and Parameter Tuning in the Context of Anomaly Detection

For our third experiment we considered Thyroid Disease dataset, which has 3772 training instances and 3428 testing instances. It has 15 categorical and 6 real attributes. The problem is to determine whether a patient referred to the clinic is hypothyroid. Therefore three classes are built: normal (not hypothyroid), hyper function and subnormal functioning. For outlier detection, 3772 training instances are used, with only 6 real attributes. The hyper function class is treated as outlier class and other two classes are inliers, because hyper function is a clear minority class.

To find the appropriate algorithm for this particular dataset, the first question is about the nature of the problem. Because it is clear that we have a classification problem (classify examples as outliers or inliers), we can reduce the search space and go to the next step, to analyze the nature of the data. As we specified in the description of the thyroid disease dataset, we have labeled data, so we can step further and analyze the dimension of our dataset. The dataset contains 3772 instances, which is less than 100K, so we can try out the SVM algorithm.

After we found the appropriate algorithm, the next step is to use our platform to minimize the error and to maximize the precision, recall, accuracy and other evaluation results. We run the algorithm multiple times. The results are presented in Table 3. Firstly, we used C = 1 and Gamma = 0.1 with rbf kernel. The results are not too promising, because having a really small recall, much of the outliers are omitted. In the next step we increased both C and Gamma, which gave us a much better results. From 0.08 the recall increased to 0.6. Even if the precision has dropped, the overall result is much higher, the model behaves much better. As we keep increasing the values of C and Gamma, we are getting better and better results until we reach a threshold, when the model starts to overfit. The best results were obtained using C = 100 and Gamma = 10.

The algorithm selection and tuning process ca be repeated if the final results are not satisfactory.

7 Conclusion

In this paper we presented a platform that implements the main functionalities needed by a developer in the process of finding the best strategy for a given domain that assures high quality problem solving in a reasonable execution time, increasing the productivity of the developers, increasing the probability of success and decreasing the costs and risks. It includes multiple data acquisition, preprocessing, anomaly detection and visualization functionalities that may be combined in a specific processing flow.

Through a number of examples we demonstrated the benefits of our algorithm selection methodology and the usefulness of our platform in finding the correct parameters for the chosen algorithm. These experiments clearly show how can we reduce the search space when a developer have to choose and adjust a proper AI algorithm.

The experimental part of the paper demonstrates some of the functionalities of the platform, including the possibility to compare the result obtained with different methods, to run the training and evaluation process multiple times, to automatically tune the parameters of the selected algorithms and to visualize the data and the results in a meaningful way.

References

1. Chandola, V., Banerjee, A., Kumar, V.: Anomaly detection: a survey. ACM computing surveys (CSUR) **41**(3), 15 (2009)
2. Estevez-Tapiador, J. M., Garcia-Teodoro, P., Diaz-Verdejo, J. E., Anomaly detection methods in wired networks: a survey and taxonomy, Computer Communications **27**, 15 (2004, October)
3. Rassam, M.A., Zainal, A., Maarof, M.A.: Advancements of data anomaly detection research in wireless sensor networks: a survey and open issues. Sensors **13**, 10087–10122 (2013)
4. Gupta, M., Gao, J., Aggarwal, C.C., Han, J.: Outlier detection for temporal data: a survey. IEEE Trans. Knowl. Data Eng. **25**(1), (2014, January)
5. Agrawal, S., Agrawal, J.: Survey on anomaly detection using data mining techniques, 19th International Conference on Knowledge Based and Intelligent Information and Engineering Systems, ed. Elsevier, Procedia Computer Science **60**, 708–713 (2015)
6. Hodge, V.J., Austin, J.: A survey of outlier detection methodologies. Kluwer Academic Publishers (2004)
7. Zareapoor, M., Seeja, K.R., Alam, M.A.: Analysis on credit card fraud detection techniques: based on certain design criteria. Int. J. Comput. Appl. **52**(3), (2012)
8. Dasgupta, A., Nath.: A classification of machine learning algorithms. Int. J. Innov. Res. Adv. **3**(03), (2016)
9. Maglogiannis, I., Karpouzis, K., Wallace, M., Soldatos, J.: Real word AI systems with applications in eHealth, HCI, information retrieval and pervasive technologies. In: Proceedings of the: Conference on Emerging Artificial Intelligence Applications in Computer Engineering, 2007. The Netherlands, The Netherlands, Amsterdam (2007)
10. Uyanık, G.K., Güler, N.: A study on multiple linear regression analysis. Procedia-Soc. Behav. Sci. **106**, 234–240 (2013)
11. Google Cloud AutoML, https://cloud.google.com/automl/

12. Kotthoff, L., Thornton, C., Hoos, H.H., Hutter, F., Leyton-Brown, K.: Auto-WEKA 2.0: automatic model selection and hyperparameter optimization in WEKA. J. Mach. Learn. Res. **18**(1), 826–830 (2017)
13. Wolpert, D.H., Macready, W.G.: No free lunch theorems for optimization. IEEE transactions on evolutionary computation **1**(1), 67–82 (1997)
14. Czako, Z., Sebestyen, G., Hangan, A.: Evaluation platform for artificial intelligence algorithms. In: Proceedings of the 10th International Joint Conference on Computational Intelligence, Vol. 1: IJCCI, ISBN 978-989-758-327-8, pp. 39–46
15. UCI Machine Learning Repository, https://archive.ics.uci.edu/ml/index.php
16. SciKit-Learn, https://scikit-learn.org/stable/

AI Turning Points and the Road Ahead

Lito Perez Cruz and David Treisman

Abstract Today Artificial Intelligence (AI) is enjoying a revival of interest. A decade ago, computer product makers avoided the term, for fear of being branded wide-eyed dreamers. Now seen in a positive light, AI is enjoying frenzied attention from the public. This is a marked change. In this paper we track and examine what led to the present day preoccupation of the public to things AI, a significant deviation from some thirty years ago. In this study, we extend the work *Making AI Great Again* where we will provide more arguments for the factors that led to this turn of events and evaluate if what we are seeing is "real" AI. Along the way, we will offer ways in keeping the AI hope real despite the sometimes exaggerated hype that could cloud the AI achievements of the past. It is our aim that this work helps in someway to prevent AI from experiencing another Winter.

Keywords Artificial intelligence · Deep learning · Agents · Ontologies · Common sense reasoning

1 Introduction

Some ten years ago, Artificial Intelligence (AI) practitioners avoided describing their work and products with the AI label. This is because many years ago, after receiving much positive publicity, AI fell into disrepute with many practitioners in the field being derided as misguided technologists. Things have certainly changed. Today AI researchers are no longer shy about their work and in fact, many more people from the public want to get into the field feeling that if they do not, they will be left behind.

L. Perez Cruz (✉)
Sleekersoft P/L, Melbourne, Australia
e-mail: lpc@sleekersoft.com

D. Treisman
CF Gauss & Associates, Melbourne, Australia
e-mail: david.treisman@cfgauss-associates.com

C. Sabourin et al. (eds.), *Computational Intelligence*, Studies in Computational Intelligence 893, https://doi.org/10.1007/978-3-030-64731-5_5

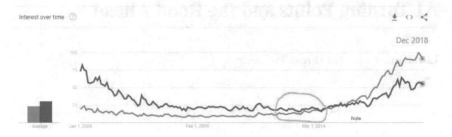

Fig. 1 The rise of AI searches in Google

Figure 1 is proof of the rise of interest in things AI which begun roughly in 2012 and distinguishes AI (the red line) from Machine Learning (ML) (the blue line).

Governments have not been immune to the wide spread interest things AI. For example, the European Union through the European Commission has agreed to boost AI investments.[1] A mere 30 years ago, such risk taking would have been unheard of. Even the UK has announced that it will be allocating 1 billion pounds to finance at least 1,000 government supported PhD research studies[2] in things AI. Similarly, the French government has recently announced its ambitious strategy to turn Paris into a global AI hub.[3]

This is a strong contrast to 1973, when the UK government released its so called Lighthill report [21] in which it questioned the soundness of the AI endeavor, suggesting it to be an illusory mirage for failing to deliver on its lofty objectives. AI practitioners, it said, failed to deliver on their promises. The report was skeptical of AI's ability to deliver solutions to real world problems. Much of the lack of usefulness of AI in practical domains was claimed in the report to be due to the inability of the techniques in scaling up to practical tasks. Years after the release of the Lighthill report decline in AI funding occurred not only in the UK but all over the world. In the era following the Lighthill report, many AI scientists and practitioners experienced trauma and shunned the use of the label AI. Nonetheless a few practitioners kept the faith of the prospects of AI. This era was so significant to the field of AI that computer science historians call the decline of AI funding and its related interest as the "AI Winter". Subsequently, they called AI's recent rise of popularity into what we have today as the "AI Spring" [11].

In this paper, we will follow the path that led to the present rise of AI activity. We approach it from the start when the visionaries of AI introduced the term and analyze what they meant by it. We will show that the present success is coming from the Deep Learning(DL) method, a technique which is founded on Artificial Neural Networks (ANN), which is itself a small aspect of ML, and thus also of AI. We will show that

---

[1] http://europa.eu/rapid/press-release_IP-18-3362_en.htm.

[2] https://www.gov.uk/government/news/tech-sector-backs-british-ai-industry-with-multi-million-pound-investment-2.

[3] https://techcrunch.com/2018/03/29/france-wants-to-become-an-artificial-intelligence-hub/.

in the majority of cases this is what people mean when they name-drop the AI label. We will show that a part of AI, which is DL, is being used to refer to the whole of AI. We then evaluate if such approaches to AI are "real AI" as originally conceived by the founders of the field. Finally we offer some lessons we can learn from the several Winters and Springs that AI has experienced in the past in the hope that AI research and development stays fresh and sustainable for many years to come.

2   The AI: Its Origin

The trouble with AI is it's defined however anybody wants to
— Martin Kihn, VP Research, Gartner

Decades ago, AI-based product manufacturers avoided the use of the term for their products, yet now, people appropriate the label with no qualms at all. Why the change? Volumes of work document the history of AI, below is a small review of interesting historical points relevant for this analysis.

The question of intelligence is really an elusive concept. There is no precise definition of intelligence and for philosophers, psychologists and educators are still trying to settle the right meaning of the term. Our understanding of intelligence is purely intuitive, when we use the word as a whole, we all get what is meant by that word. We often have a notion of intelligence, yet it continues to be vague and conceptualizing it into words is easier said than done. This is demonstrated by a simple dictionary definition of "intelligence"—"the ability to acquire and apply knowledge and skills". This definition applying to human intelligence is in its own right is too broad if not too ambiguous, and points to the ease of proliferation of the use of the AI term. Because of this imprecision in identifying human intelligence, we face the same dilemma when it comes to machine intelligence, i.e., AI. It stands to reason that if we cannot even settle the meaning of intelligence in humans, how can we try to use it to judge machine intelligence? It is on a similar basis that Alan Turing asked if machines could think. In this context, when computer scientists say "think", they mean the ability to make deductions or produce more knowledge from existing information.

Similarly, when a program makes predictions with great accuracy, is this intelligence? Another example would be when a computer program performs optimization, is this intelligence? What about if the program is able to accurately classify an image of an object. No doubt in these situations, and from a human standpoint, seeing a machine having such abilities will cause people to marvel with awe. If something is automated, is that a demonstration of its capacity to think? These critical questions have been central to AI researchers but from the layman's point of view, they are good enough. Experts recognize the nebulous issue of intelligence so much so that in such inaccuracy they rally for a more formal and accurate definition of "intelligence" [28]. This ambiguousness, we believe, is a source of confusion when AI researchers see the term used today [14] [12]. Critics of the present AI popularity are quick to point out this problematic issue [12].

To shed better light on what is meant by AI, we are guided by Carl Sagan's famous rule: "You have to know the past to understand the present". We will apply this rule in order to get a better understanding of what is the current period of public awareness of all things AI.

The man responsible for coining the phrase "Artificial Intelligence" (AI) is John McCarthy, the inventor of the LISP programming language. McCarthy devised this label in 1956 at a Darthmouth College conference attended by personalities such as Marvin Minsky, Claude Shannon and Nathaniel Rochester and another seven others of academic and industrial backgrounds [5, 27]. These researchers were organized to study if learning or intelligence (as stated in their purpose for meeting),

> ...can be precisely so described that a machine can be made to simulate it. An attempt will
> be made to find how to make machines use language, form abstractions and concepts, solve
> kinds of problems now reserved for humans, and improve themselves. [27]

At that conference, Allen Newell and Herbert Simon with J. Clifford Shaw of Carnegie Mellon University showcased their Logic Theorist program [15, 27] with thunderous results. This program was a reasoner and was able to prove most of the theorems in Chap. 2 of *Principia Mathematica* by Bertrand Russell and Alfred North Whitehead. The computer scientists in attendance came from strong mathematical backgrounds like the field of foundations of mathematics. Therefore they welcomed this result for they appreciated how difficult it was for a computer to have such a skill which mathematicians take great pains to accomplish. As a result many believed that all contemporary mathematical theories could be so derived in such a way. Ironically they tried to publish their work at the *Journal of Symbolic Logic* but the editors rejected it, on the basis that it was a machine that did it and not a living mathematician. Apparently they took it for granted that a computer did all the derivation of the theorems, which to those who are in-the-know would have been amazed.

Although the AI term was coined in 1956, its conceptualization can be traced back further to the work done by Warren McCulloch and Walter Pitts in the area of computational neuroscience [27]. Their 1943 work entitled *A Logical Calculus of Ideas Immanent in Nervous Activity* [15, 25, 27] proposed a model for artificial neurons as switches with "on" and "off" states. These states are seen as equivalent to a proposition for neuron stimulation. McCulloch and Pitts showed that any computable function along with the logical connectives (and, or, not, etc.) can be computed by some network of neurons. The interesting part is that McCulloch and Pitts suggested these artificial neurons could learn. In 1950, Marvin Minsky and Dean Edmonds inspired by this research on computational neural networks (NN), built the first hardware-based NN computer. Minsky later would prove theorems on the limitations of NN [27].

We can see here that there are twin developments in the history of modern AI as we know it and we shall come to this in the sections below. The 1956 first definition of AI relied on logical theories while the 1943 conceptualization as we will see relied on a graph or network theoretical framework. In all of these, the most notable phenomenon is the high optimistic pronouncements that emerged right after some remarkable breakthroughs had occurred in what a program or a machine could do. We will see that these types of hopeful pronouncements can easily turn to hype and will bear upon our analysis later on in this work.

3   AI Subcategories

In this section we discuss various angles and categories of viewing AI methods. These types or views of AI are almost never distinguished in lay conversations and the media hardly knows the existence of these ways of looking at AI. It will be helpful to deal with these distinctions, for by doing so, we can appreciate the rise of the hype in AI and understand where we might look for hope.

3.1   Symbolic Versus CI

In this portion we will analyze the methods in achieving the goal of AI.

In Sect. 2, we mentioned an AI group organized through McCarthy. This group, we notice, rallied around the result produced by the team of Newell-Simon-Shaw whose result was based on the use of logic. Because of this, the group gathered by McCarthy proceeded to work on the use of logic in AI. This approach models intelligence formally as equivalent to the activity of logical reasoning. Since thinking involves logic, they hoped that through the use of formal logic, they would reach their goal. This approach has consequently been called by some as the *Symbolic* approach to AI and was popular in computer science. Authors often call this approach Good Old Fashion AI (GOFAI). Most of these people apart from Minsky, worked on this field and for a while gathered momentum primarily because it was programming language based and due to the influence of Newell and Simon's results [11].

Those working on NN like McCulloch and Pitts were called *Connectionists* since by the nature of networks, must be connected. However, contemporary researchers now view neural networks as a subgroup of a much larger body of approaches antithetical to the symbologists, which is now called *Computational Intelligence*(CI) [15]. Flasinski [15] lists two more sub-methods along with the connectionists approach— namely the *mathematics based* method like fuzzy sets and fuzzy logic and the other one being the *biology based* method such as evolutionary and genetic programming etc. The Symbolic and Connectionist groups continue to debate with each other on the proper method for addressing the challenges facing AI [11, 32].

The evolution of these two major groups contribution to modern AI may be depicted in Fig. 2.

Fig. 2 AI's modern
development

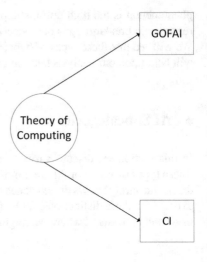

It should be noted that almost no mention is said about these types of AI groups
or approaches when the AI label is used in conversation. Mentioning the type of AI
approach taken would make clear what is at play when someone especially describes
AI type results in glowing colors.

3.2 Narrow, Weak or Broad, Strong

We turn now to classify the degree of achievements of those working in the field of
AI.

There is a belief that what the human mind can do, the computer should be able
to do as well. This belief is founded on the theory that the human brain is just like a
machine, or is also a computer. So, what it is able to do, that should be transferable to
hardware and software. This was the viewpoint taken by Newell and Simon in 1976.

However, Searle introduced the categories Strong AI versus Weak AI [29]. Sometimes Strong AI is also called Hard AI. In Weak AI, machines are considered to
only act *as if* they were intelligent [27]. In Strong AI the "brain in the box" idea,
the computer is *actually* thinking on its own in an autonomous fashion, albeit that
it is delimited by the hardware box and not just simulating the thought processes of
humans. Therefore, Strong AI implies that the computer should be able to solve any
problem and thus will grow, learn and mature on its own. Some even believe that a
computer can have its own self-consciousness. Collectively these features of strong
AI are referred to as Artificial General Intelligence (AGI) or Broad AI [11].

By contrast, in Weak AI, the computer is seen as a device that is geared up to
solve only specific or particular tasks or to mimic an aspect of human thinking.
In this regard, a computer can be skillful in some specialty but unlike its humans
counterpart, it cannot be multi-skilled. Weak AI is often referred to as Narrow AI.

By far the lack of distinction between Weak and Strong AI is the common source of misunderstanding when one hears about AI. A basic search of LinkedIn using the term "artificial intelligence" reveals how the term AI is loosely used in articles or posts with no distinction as to whether it is weak or strong. In most of these searches, AI is depicted in two extremes: either in the context of profound admiration or in terms of its apparent dangers. Somehow these extreme depictions reinforce the lack of distinction between Weak and Strong AI as ideas continue to get lost in "translation" each time the AI term is utilized [11].

4 The AI Wave

We will now track the rise, fall and the rise again of AI. It is also instructive to learn the names of certain events that happened in relation to funding of projects which participate in the search of AI. Historians say AI is in an *AI Spring*, when the investor community are enthusiastically injecting funds into organizations who are undertaking AI research. They say AI is in an *AI Winter* when investors are skeptical of AI projects and are obviously not buying into the idea of funding them. Researchers trace this terminology to Richard Gabriel, a popularizer of Common Lisp and founder of Lucid Inc, a company in the 80s that focused on Lisp Programming. Historians believe the spring and winter analogy got introduced around 1990–94. We depict this rising and falling in AI funding through Fig. 3 [11].

AI researchers from 1952–1969 were gaining the attention of the public because of the successes that they had been demonstrating. For example, Newell and Simon extended their Logic Theorist to General Problem Solver (GPS) which could solve

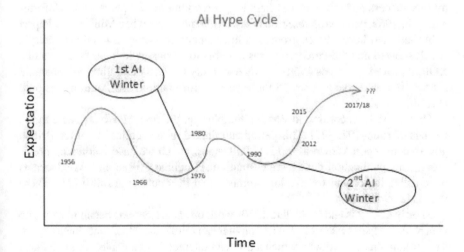

Fig. 3 AI's present hype cycle

puzzles. This program mimicked the way a typical human might proceed in solving a problem or task, such as establishing sub-goals and planning possible actions. As the then AI community were trained in mathematics based disciplines, observing a program could prove mathematical theorems was seen by them to be very impressive due to their knowledge that the tackling of similar problems by human beings involved considerable creativity, imagination and foresight. They were specially impressed when Herbert Gelernter show cased his Geometry Theorem Prover. For at that time, seeing a computer play checkers against a human being and even defeating its human opponent provided significant impetus to the hope that machine intelligence was surely just around the corner. The symbolicists AI practitioners indeed dominated this part of history and they obtained encouragement by way funds flowed to their projects. Seeing computers interact this way, does at first sight give the impression that they can think, for that is what is needed in a game like chess or checkers. In his over enthusiasm, Herbert Simon said that in 10 years (at that time), computers would be chess champions and prove significant mathematical theorems [27]. It was further predicted that computers would also learn and be creative. Needless to say it was a great let down when these predictions did not happen in the subsequent 10 year period. This was because Simon's forecast was based on a small sample of AI successes. Playing chess is much less complicated than language translation, for example. Although Simon was not wrong in terms of what the computer can achieve, his crystal ball like and optimist prediction of the time did not help the cause of AI. Nonetheless, some of these aspects did happen but this was much later, some 40–50 years from his statement [11].

We note that in parallel, the connectionists were also gaining ground with their idea of the *perceptron*, which is a single layer NN, the precursor to NN. Through the work of Frank Rosenblatt, roughly between 1957-58, the perceptron gained a following because it could recognize simple patterns. Many were were encouraged by this success and many researchers began working on the perceptron. Unfortunately, in 1969, perceptron research received a major blow when Minsky and Papert published their book "Perceptrons: An Introduction to Computational Geometry", which showed that the Perceptron was not able to learn simple logic functions like XOR, it proved the impossibility of NN to classify patterns of nonlinearly separable classes. It would take another 15 more years before faith in NN could be restored [11, 30].

On this basis, according to Russell and Norvig, the first AI Winter came in the periods of 1966–1973 [27]. This period culminated in the publication of the British government report known as the Lighthill report which we have mentioned previously. This publication demonstrated that an AI bubble existed and AI advocates were called to account for making promises that they could not fulfill [21]. There was disillusionment.

Shortly after this and in the late 1970s work on *expert systems* began to rise. The idea behind expert systems is that a computer can mimic the job of a sit-down expert, a consultant on the field with which a user can interact. For example, it can be like a doctor that helps another doctor diagnose a patient's set of symptoms. This brought marginal success to AI, since expert systems are actually demonstrations of Weak AI.

It would take symbolic AI scientists' work on *knowledge representation systems*(KR) to bring new life to AI in the early 1980s. Unfortunately, history would repeat itself as another AI Winter descended for the second time as many KR systems failed to deliver on over-hyped promises. Bizarrely, for example, KR systems promised that AI could do anything [10]. Critical of this idea, Coats pointed out that expert systems were unable to deal with incomplete or inconsistent information, and as such could not reason from first principles nor did it have common sense intuition [3]. Similarly, it failed when it came to knowledge acquisition and execution, the process of deduction took many cycles of processor execution time to get the needed answer in time [10].

What might have been the causes of the first and second AI Winters? When the computer trashes a human chess champion on a game of chess, does this imply that it can do a lot of things in a superior way to what humans can do? In those days there was something to marvel at when a human chess champion was humiliated by a mere nuts and bolts machine. But, can we now conclude from this that the computer can translate a document written in one non-English language to English? Furthermore, most of the human problems which we want AI to solve are more complicated than mere game playing. How would that capability help an ordinary citizen, how would that assist in easing the daily chores of life? These are non-trivial problems where the computer gets bogged down in dealing with the solution space required to solve intricate issues. Over promises in AI combined with difficulties in hardware and software resources spelled disaster for AI, the fall into an AI Winter was not hard to imagine.

It is not unreasonable to blame the instigation of AI Winters on the shoulders of AI proponents. Usually, when Weak AI gets attention due to some note worthy result, other AI promoters are quick to extend this success to AGI, implying that an AGI dominated world is just a few meters away. For example, the famous Andrew Ng tweeted that radiologists will soon be obsolete [26]. It seems when AI experiences success, AI enthusiasts get carried away making unbridled promises of what it can do and deliver. Though the criticism might be against AGI, it also negatively affects Weak AI as well. We see this for instance when Google's DeepMind AlphaGo[4] beat its first human opponent in the game of Go. Immediately we had experts saying it is now time to embrace AI with one expert saying AI will surpass human intelligence in *every field* and develop superintelligence [18]. We are not advocating AI phobia, not at all, but this assertion borders on tabloid fodder. Of course, a computer will beat a human in a game like Go. For one thing, humans get tired, they carry personal anxiety and family issues to the game, they could be suffering some illness as they play etc. Humans have many forces that can distract their concentration; but computers do not have any of these. So for sure, a computer will beat a human when it comes to game playing [11].

AI experts have no one but themselves to blame for the disillusionment they bring to the investor community. For when the pot for funding AI projects goes dry, they have to look to themselves for correction.

---

[4]https://deepmind.com/research/alphago/.

5 Leading the Way

5.1 Deep Learning to the Rescue

Going back to Fig. 3 [11], we see the boom and bust nature of AI funding. Some observed its comeback in 2015 [1]. Others believe it came to the mainstream in 2016 [2]. A slow reintroduction of AI back into the scene can be traced as far back as 2012. This is confirmed by Statista (Fig. 4) which reports a significant funding increase in AI that happened world wide since 2012[5] [11].

Starting from a fund of $0.5B in 2012, it jumped 10 fold to $5.0B by 2016. This is a massive increase in a 4 year period. On the same site it is estimated that in 2018 the global AI market was worth approximately $4.0B with the bulk of AI investors focused on AI applications for the enterprise [11].

What is causing this sudden AI Spring? If we go by the social media outlets and other internet-based publishing methods, we will see that *Deep Learning*(DL) is the prime driver of this apparent AI Spring [1, 2]. DL is a more network dense configuration of ANN, it is a heavily multi-layered ANN. DL has proven to be reliable when it comes to supervised learning tasks and especially to the classification task. Figure 5 is from [16] and illustrates an ANN.

Haykin [16] or Bishop [4] gives a detailed treatment of ANNs and we will use their notations here. The simplest case is ANN with two layers. The linear combination of input variables are called $x_1, x_2, ...x_D$ and going into the first hidden layer with M neurons we get the output activation

$$a_j = \sum_{i=1}^{D} w_{ji}^{(1)} x_i + w_{j0}^{(1)} \tag{1}$$

The $w_{ji}^{(1)}$ are the parameter weights and $w_{j0}^{(1)}$ are the biases with the superscript (1) designating the first hidden layer. They then get transformed by an activation function $h(\cdot)$ like $z_j = h(a_j)$. Then we get for K unit outputs the following

$$a_k = \sum_{j=1}^{M} w_{kj}^{(2)} z_j + w_{k0}^{(0)} \tag{2}$$

These then finally get fed into the last activation function y_k

$$y_k = f(a_k) \tag{3}$$

In DL, this multi-layer ANN is made more intricate with several hidden layers in between.

---

[5]https://www.statista.com/statistics/621197/worldwide-artificial-intelligence-startup-financing-history/.

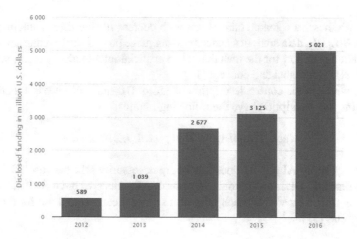

Fig. 4 AI Funding in millions by statista

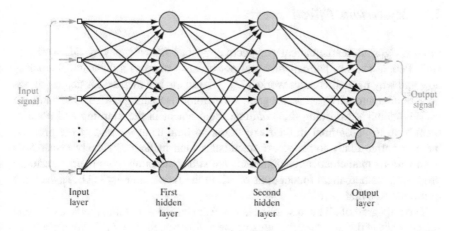

Fig. 5 Multylayer ANN

We have some evidence that when AI is mentioned, the speaker really means DL. One indicator of this behavior is to see how data science training groups are using both terms. For example in Coursera doing a cursory search on the phrase AI in its courses will not give results of courses with AI in their titles. Instead, what one gets back are several courses on DL and machine learning(ML)! This suggests that AI and DL are being made to correlate with only one result having the word AI in it and it is IBM's "Applied AI with DL". This shows that the term AI has been made synonymous in common parlance with that of DL and provides the reason why those DL and ML courses come up [11].

The search result is almost the same in edX but even more telling. In edX, the search came back with 6 courses having the title artificial intelligence in them with two from Columbia University and four from Microsoft. The interesting part is that along with these, the search came back with courses having titles containing data science, ML, DL, data analytics and natural language processing to name a few. Here we see strong support for the idea that when artificial intelligence is mentioned, DL (and ANN) is an associated concept [11].

Somewhere in the course description of ML or DL subjects there is a mention of AI making the description have the following implication:

$$deep\_learning \Rightarrow artificial\_intelligence$$

DL may imply AI but AI does not imply necessarily DL, because AI is much broader than DL. DL is an approach to some of AI's concerns but it does not address the whole of it. Some have much faith in it saying it can make a run for the whole AI enterprise [31].

5.2 Mysterious Effectiveness

From a rigorous mathematical perspective, no one knows yet why DL works so well. People use DL heuristically and not because one has a clear understanding how and why its mathematics works [22]. A heuristic is a method for getting things done or an approach to achieve a goal. An example is making coffee. One can have various methods for doing this common requirement in the morning and none of them have to be optimal in their execution, so long as at the end of the process, there is coffee ready to drink. AI scientists which include obviously researchers from various mathematical backgrounds are still abstracting from their experience proposing mathematical foundations of DL to the AI community [11]. An example of this work is that of [6].

The holy grail of all the disciplines, be it from business, natural, medical or social sciences are in the quest for that elusive the f function in $y = f(x)$. We are always in search to find such a function for in succeeding in its discovery, human life's difficulties might be minimized; we get less controlled by the elements around us, we get a bit of knowledge and adjust our decisions that will yield maximum benefits. This does not stem from DL, but simply an obvious human need to anticipate and predict an outcome. DL and ANN come close to approximating this f [11].

We may generalize DL formally in the following manner [11] by taking our cue from [6]. Assume we have L layers in a DL. Let X_i be inputs coming to neurons at layer i, and let W_i be the weights at layer i. We will express the whole DL as a composition of functional transformations $f_i : X_i \times W_i \longrightarrow X_{i+1}$, where we understand that X_i, W_i, X_{i+1} are inner product spaces $\forall i \in L$. Further let x be a vector and $x \in X_1$. Then we can express the output function produced by the DL as

$$\widehat{F} : X_1 \times (W_1 \times W_2 \times \cdots \times W_L) \longrightarrow X_{L+1} \tag{4}$$

We purposely use the 'hat' notation to emphasize the fact that it is estimating the real F behind the phenomenon we are modeling.

If we understand further that f_i as a function depending also on $w_i \in W_i$ then we can understand $\widehat{F}$ as

$$\widehat{F}(x; w) = (f_L \circ f_{L-1} \cdots \circ f_1)(x) \tag{5}$$

We come now to a very important result [20].

Theorem 1 (Hornik's Theorem.) *Let F be a continuous function on a bounded subset of n-dimensional space. Let ε be a fixed tolerance error. Then there exists a two-layer neural network $\widehat{F}$ with a finite number of hidden units that approximate F arbitrarily well. Namely, for all x in the domain of F, we have $|F(x) - \widehat{F}(x; w)| < \varepsilon$. [17].*

This is where the notion that DL is a *Universal Approximator* comes from. We can set this ε as small as we like and still find the $\widehat{F}(x; w)$ for this $F(x)$. Further, F can be approximated by a single hidden layer ANN [11].

Because of this, it is no wonder that when it comes to ML data scientists often go first to DL to save them. It is their "first stop" of choice and only when it fails to give the precision they want do they quit DL and try something else. From a logical or conceptual standpoint, a simple ANN will do. However, DL, historically, is a re-branding of the work of Hinton on neural networks, who even shied away from using the term for describing his doctoral research [11, 31].

Having said the above, DL along with other ML techniques have had successes due to conditions that were not present more than 30 years ago which are (a) the availability of massive amounts of data which help in the statistical process, (b) growth in the maturity and efficiency of algorithms, and (c) availability of faster and cheaper computers accessible easily to a great number of DL practitioners. These factors together overcome the disabling conditions that were present when lofty pronouncements in AI were being made three decades ago.

6 Agents and Information Architectures

Lay people are not aware of the view of AI inherited from GOFAI as it does not get media attention. Under this view AI programs are embodied as *rational agents* [27] [28] which in itself date far back in 1990s [11].

This rational agents view say that we have AI in a program if it can do *bounded optimality*, i.e., "the capacity to generate maximally successful behavior given the available information and computational resources" [28]. Thus, it has AI if it will choose the best course of action under the circumstance. The operative word is *best action*, and it is based on a key performance index, a success indicator that is measurable and part of the program. Under this view, AI actions in answering questions are "passive" as it performs no actions no more than a calculator. If the focus is on prediction, commonly found today, but no automatic action arises from it, then it falls short of AI. Most ML systems are like this, they predict or cluster, so this view begs that these systems must be wrapped in a behavioral program which acts upon those ML outcomes. Accordingly, the product is operating as a "consultant" and is just a special case of the high level action oriented function performed by an agent. We are not saying there are no recommendation systems out there which may be viewed as an action. However, by the way the AI term is used, this is not happening pervasively in the community [11]. Similarly, how the AI term is employed by the greater community, such action oriented notions are left out of the discussion [11].

Formally, an agent g transforms or maps a series of observations $\mathcal{O}^*$ to a set of actions $\mathcal{A}$.

$$g : \mathcal{O}^* \longrightarrow \mathcal{A} \tag{6}$$

Viewed AI this way, then it is easy to assess whether or not a so called entity is doing AI by simply examining if the said entity transforms what it senses into behavioral actions [27].

The main proponent of this view is Stuart Russell (University of California, Berkeley) and Peter Norvig (Director of Research, Google Inc.) Their textbook *Artificial Intelligence: A Modern Approach* [27] adopts such a view. The textbook is found in universities in 110 countries and is the 22nd most cited book in Citeseer.[6] In all likelihood a computer science graduate would have come across this view of AI.

In 2016, and at the time of writing, the AI label has been the buzz of both the academe and the industry. Observers have begun to come out of the wood work to give a more tempered hold on the hyperbole surrounding AI. Early [14] wrote a sober paper called *No AI without IA*. Here IA stood for Information Architecture (IA), also known as an ontology or KR. Simply put, an ontology stores knowledge of a domain of interest which forms the foundation of computer reasoning wherein agents can interrogate and formulate the next action to choose. This view believes it will be hard for an AI product to run autonomously if it has no reasoning capability so that it can react rationally against its ever changing live environment. He further asserts that ontologies capture "common sense" information, it captures context where content is found. This is needed when we want to move further forward from prediction, some structure of data has to be engineered in an AI product. Without this, according to Early [14], no efficient personalization which AIs should provide can be achieved from a customer service standpoint. Historically, ontologies come from the efforts of computer scientist in the GOFAI community.

None of the media outlets have questioned claimers of AI as to whether or not, an IA is part of their AI offering or is it embodied in an agent like program. We have seen no prolific mentioning of this in the press.

---

[6]http://aima.cs.berkeley.edu/.

7 Real AI and Lessons Learned

7.1 Common Sense Reasoning

If we recall in Sect. 2, we made mention of the statement of objective of the 1956 meeting when McCarthy and his colleagues latched on to the AI phraseology. The objective has two parts. Many focus on part a) of the quote which is about simulating intelligence, but part b.), the part that says it is able to use language and form abstractions and concepts is an elaboration of part a.). The latter part defines what is meant by simulated human intelligence.

Hector Levesque [19] argued that McCarthy in 1956 was advocating the idea of common sense present in a program, if it is to be worthy of being called having AI. Levesque quotes McCarthy [19] here:

> We shall therefore say that a program has common sense if it automatically deduces for itself a sufficiently wide class of immediate consequences of anything it is told and what it already knows.

In DL you have to train your ANN configuration using lots of samples of data, lots of them and with a tremendous effort in labeling data. Human beings are not like this. Humans do not need legions of examples to learn. A child can be taught by one example, we say "this thing we see, which is round, and yellow is called a lemon, so say lemon", and the child repeats you and she gets it. There is something about us humans that we are by default able to detect the form, shape, shade etc. of an object, so much so that when you present to child, a slightly brown lemon, she can still detect that the object is still a lemon.

The above example shows that the McCarthy-Levesque theory about common sense knowledge makes a great point in that indeed, something is intelligent if it is able to apply previous knowledge to the present situation and act accordingly. In this sense DL by itself, though by no means do we deny its utilitarian benefits, lacks common sense.

Similarly, Wiggers [34] reported that Geoffrey Hinton and Demis Hassabis, both well known personalities in the field of DL with Hinton being considered the "father of DL", as saying that while DL is providing successes, AGI is no where close to being a reality. Levesque [19], and McCarthy if he were alive, would have applauded such humility, both have argued for the restraining of the AI hype.

7.2 Insights from Economics

In a way the resurgence of AI has been driven by its contemporary use in the enterprise application market whereby computer software is implemented to satisfy the needs of the organization. Much of this software relies on the market's understanding of

DL. Which, in turn, raises concerns as to where it is in the current phase of Gartner's hype cycle.

Unsurprisingly, functional business areas that contribute the greatest value to firms are the ones to provide the greatest gains through the adoption of AI. The oft cited examples are that of personalized sales and marketing efforts in retail companies and predictive modeling of supply chain or manufacturing processes.

McKinsey estimates that these two functional areas make up more than two-thirds of AI opportunity. McKinsey anticipates that AI can create $1.4 to $2.6 trillion of value through its deployment in sales and marketing functions and $1.2 to $2 trillion in supply chain and manufacturing. The latter having significant value contribution through predictive maintenance—the analysis of productive efficiency of machines using historical performance data to predict maintenance shut downs and other delays [7, 13].

The estimated value creation of AI in Chui et al.'s [7, 8] studies are primarily based on DL techniques that use ANN. Admittedly, this is recognized in these studies to be a narrower component of the broader AI field—what we refer to as weak AI.

Similarly, Chua et al. [7] recognized that data demands in terms of volume and quality together with organizational adoption of technology and skills are the greatest challenges faced in unlocking the potential of DL. With the synonymous and often misunderstood use of weak AI techniques with the promise of hard AI technology lies the risk of realizing an AI Winter.

In order to address this concern, the relationship of DL with that of AI can be expressed in terms of the economic model of supply and demand. In terms of this model, DL through its real world usage for prediction and refinement in practice can be seen as the outcome of technology production, or supply. Whereas, the demand for DL stems from the use of AI in enterprise applications. The intersection between AI and DL is the equilibrium price between supply and demand and can be interpreted as the relative value ascribed to AI by enterprises.

This is where the distinction between strong and weak AI becomes fundamental to understanding the current phase of Gartner's hype cycle and whether another AI winter is likely to arise. DL by all accounts is, in fact, weak AI as it is geared to solve specific or particular tasks, namely prediction. However, AI, in its true sense, is expected to be hard AI as it maintains a capacity of intelligence to solve any problem. Thus the relationship between supply and demand can be updated to be the relationship between the supply of weak AI with the demand of hard AI.

This distinction matters as the predominant AI technique applied in enterprise applications is DL. Like all forms of technology in business, DL will suffer from diminishing marginal returns. This implies that as more and more DL is applied the lower there usefulness in fulfilling the needs of the enterprise application market.

Without the development of hard AI, as some point the usefulness of soft AI will reach its limitation and, in turn, be supplied in excess relative to its demands. This will create a surplus of soft AI. To return to a point of equilibrium a downward correction of the value of AI will occur, thereby triggering a new AI winter. On this basis DL cannot be allowed to be the only form of AI.

7.3 Keep the Hope Alive

Companies like the McKinsey Company [9] are forecasting a glowing future for AI, saying that it will have an impact worth $T 3.5–15.4. Is this fuel for the hype, don't we have reason to fear that another backlash Lighthill Report [21] might be just around the corner? Peikniewski [26] one of the main attackers of the AI hype, along with people like Gary Marcus [23], thinks the AI winter has already begun. Peikniewski has not deviated from his position, persistently blogging about it even as we write. One of the cases Peikniewski shoots down is the AI present in the Tesla Full-Self-Driving car. We submit that it is not productive for computer scientists and AI researchers that investors look at things AI as some form of marketing stunt. Having an AI Winter is no good to all of society [11].

What should we do keep the hope from becoming hype? We suggest consideration of the following [11]:

1. Follow the late John McCarthy, who said "it would be a great relief to the rest of the workers in AI if the inventors of new general formalisms would express their hopes in a more guarded form than has sometimes been the case" [24]. It is indeed a very wise counsel.
2. Never promote the achievements of weak AI into strong AI. It seems when weak AI succeeds, the enthusiasts are quick the extrapolate this to the immediate possibility of AGI. Part of this is to educate the media on such distinctions. It is not productive to go along their sensational spin without saying anything. The continuous use of Strong AI promotion for Weak AI results will drive demand for Strong AI. This does not set the right expectation, when in fact what the industry is able to supply is Weak AI. Thus a disconnect will happen and no doubt the decline in investors' returns will lead again to another AI Winter.
3. ML and DL practitioners are not often aware of other parts of AI. For example, some of them got onto the bandwagon and are oblivious of the existence of symbolic AI or the presence of agents and ontologies etc. Such awareness may help them in not feeding the troll of AI hypers.
4. Realize that DL though useful for certain tasks is not enough in most cases. This is because prediction does not address subsequent actions that are needed after that. The next best action to take involves reasoning beyond what DL provides. Many believe the step forward is to combine symbolic AI with connectionist AI [33]. We also concur with [23] which says that DL is not adequate to produce higher level of intelligence. We believe that the best way to do this is embedding DL into some form of an agent combined with ontologies. By economic reasoning, DL will not be the sole source of AI success. This is not sustainable in the long-run.
5. A balanced message is to communicate that we are not there yet, such as what this paper is doing in sobering up the community. We need to adjust the expectation of investors. Promising modest jumps in AI ability then exceeding it is a far more suitable outcome than raising the promised bar and then failing to clear it. The latter brings disrepute and makes us worst off.

8 Conclusion

In this paper, we investigated the path that led to the present rise of AI activity trying to understand the commotion we are observing in contemporary academe, media and business. It is not an exaggeration that all things AI, purported or otherwise, are currently approached and reported with much intense enthusiasm and excitement.

In our work, we clarified several nuances of AI dealing with its categories and methods of approach. Indeed, we share the evaluation of the many that DL is currently spearheading the rise of things called AI and that the AI term is thrown around to describe activities involving mainly DL. It is especially effective for tasks that involve classification and prediction. However, AI as originally conceived goes beyond prediction and classification. In fact, if we pit what is happening in the field against the original understanding of AI, we have to conclude that DL is not yet AI. Common sense reasoning is not yet available in products and systems out there. Further, DL is a piece of the whole of AI, but it is a part and not the whole of AI. There are other pieces of AI that need to come together to build a whole AI product as envisioned by the originators of the concept and the term. For example, DL and other ML techniques should be combined and embedded in an agent technology that uses ontologies for effective reasoning and deduction.

In our observation, often, AI hope turns and deteriorates to AI hype. We did not leave with dead end critique of the hype, rather, we provided the way forward for keeping the AI hope alive. It is our wish that the AI hope remain and is sustained by guarded enthusiasm for many years to come.

Disclaimer

The views and opinions expressed in this article are those of the authors and do not necessarily reflect that of their past or present employers.

References

1. Ahvenainen, J.: Artificial Intelligence Makes a Comeback. https://disruptiveviews.com/artificial-intelligence-makes-a-comeback/ (Nov 2015)
2. Aube, T.: AI and the End of Truth—The Startup—Medium. https://medium.com/swlh/ai-and-the-end-of-truth-9a42675de18 (Feb 2017)
3. Bell, M.: Why expert systems fail. J. Oper. Res. Soc. **36**(7) (July 985)
4. Bishop, C.: Pattern Recognition and Machine Learning. Springer (2006)
5. Buchanan, B.G.: A (Very) Brief history of artificial intelligence. AI Mag. Mag. **26**, (2006)
6. Caterini, A.L., Chang, D.E.: Deep Neural Networks in a Mathematical Framework. Springer (2018)
7. Chui, M., Henke, N., Miremadi, M.: Most of AI's Business Uses Will Be in Two Areas. Harvard Business Review (July 2018)

8. Chui, M., Manyika, J., Miremadi, M., Henke, N., Chung, R., Nel, P., Malhotra, S.: Notes from the AI Frontier: Applications and Value of Deep Learning (April 2018), https://www.mckinsey.com/featured-insights/artificial-intelligence/notes-from-the-ai-frontier-applications-and-value-of-deep-learning
9. Chui, M., Manyika, J., Miremadi, M., Henke, N., Chung, R., Nel, P., Malhotra, S.: Notes from the AI Frontier: Insights From Hundreds of Use Cases. McKinsey Global Institute (2018)
10. Coats, P.K.: Why expert systems fail. Financ. Manag. **17**(3) (Autumn 1988)
11. Cruz, L.P., Treisman, D.: Making AI great again. In: Madani, K., Warwick, K. (eds.) 10th International Joint Conference on Computational Intelligence. SCI Press (2018)
12. Datta, S.: The Elusive Quest for Intelligence in Artificial Intelligence. http://hdl.handle.net/1721.1/108000 (2017), http://hdl.handle.net/1721.1/108000, http://hdl.handle.net/1721.1/108000
13. Dilda, V., Mori, L., Noterdaeme, O., Schmitz, C.: Manufacturing: Analytics unleashes productivity and profitability (August 2017), https://www.mckinsey.com/business-functions/operations/our-insights/manufacturing-analytics-unleashes-productivity-and-profitability
14. Earley, S.: No AI without IA. IEEE IT Prof. **18** (2016)
15. Flasinski, M.: Introduction to Artificial Intelligence. Springer (2016)
16. Haykin, S.: Neural Networks and Learning Machines, 3/e. Pearson (2008)
17. Hornik, K.: Approximation capabilities of multilayer feed forward networks. Neural Netw. **4**, 251–257 (1991)
18. Lee, N.: AlphaGo's China Showdown: Why it is time to embrace Artificial Intelligence. Sothern China Morning Poast—This Week In Asia (2017), https://www.scmp.com/week-asia/society/article/2094870/alphagos-china-showdown-why-its-time-embrace-artificial
19. Levesque, H.J.: Common Sense, the Turing Test, and the Quest for Real AI. MIT Press (2017)
20. Lewis, N.D.: Neural Networks for Time Series Forecasting in R. Lewis, N. D (2017)
21. Lighthill, J.: Artificial Intelligence: A General Survey (1973)
22. Lin, H., Tegmark, M., Rolnick, D.: Why does deep and cheap learning work so well? J. Statist. Phys. **168** (2017)
23. Marcus, G.: Deep learning: a critical appraisal. arXiv:1801.00631 (01 2018)
24. McCarthy, J.: Review of Artificial Intelligence: A General Survey (Jun 2000), http://www-formal.stanford.edu/jmc/reviews/lighthill/lighthill.html
25. McCulloch, W., Pitts, W.: A Logical Calculus of Ideas Immanent in Nervous Activity. Bulletin of Mathematical Biophysics pp. 115–133 (1943)
26. Piekniewski, F.: AI Winter Is Well On Its Way. https://blog.piekniewski.info/2018/10/29/ai-winter-update/ (2018)
27. Russel, S., Norwig, P.: Artificial Intelligence: A Modern Approach, 3rd edn. Prentice Hall (2010)
28. Russell, S.: Rationality and intelligence: A brief update. In: Muller, V.C. (ed.) Fundamental Issues of Artificial Intelligence. Springer (2016)
29. Searle, J.R.: Minds, brains, and programs. Behav. Brain Sci. **3**, 417–424 (1980)
30. da Silva, I.N., Spatti, D.H., Flauzino, R.A., Liboni, L.H.B., do Reis Alves, S.F.: Artificial Neural Networks. Springer International Publishing Switzerland (2017)
31. Skansi, S.: Introduction to Deep Learning. Springer International Publishing AG (2018)
32. Smolensky, P.: Connectionist AI, symbolic AI, and the brain. Artif. Intell. Rev. **1**, 95–109 (1987)
33. Sun, R., Alexandre, F.: Connectionist-Symbolic Integration: From Unified to Hybrid Approaches. L. Erlbaum Associates Inc (1997)
34. Wiggers, K.: Geoffrey hinton and demis hassabis: Agi is nowhere close to being a reality. https://venturebeat.com/2018/12/17/geoffrey-hinton-and-demis-hassabis-agi-is-nowhere-close-to-being-a-reality/

Efficient Approximation of a Recursive Growing Neural Gas

Jochen Kerdels and Gabriele Peters

Abstract The recursive growing neural gas algorithm (RGNG) is a variant of the classic GNG that was specifically designed to model the response behavior of groups of biological neurons. It was used successfully to describe the behavior of entorhinal grid cells as well as entorhinal cells that show grid like activity in response to saccadic eye movements. More recently, the RGNG algorithm was integrated into a model of cortical column function as part of an autoassociative memory cell. To facilitate future research involving the simulation of hundreds to thousands of neuron groups we present an alternative algorithm to the RGNG as a drop-in replacement in the context of neuron group modeling. The *differential growing neural gas* (DGNG) is structurally less complex, more efficient to compute, and more robust in terms of the input space representation that is learned while retaining most of the RGNG's important characteristics. We provide a formal definition of the DGNG algorithm and demonstrate its characteristics with a first set of experiments.

Keywords Recursive growing neural gas · Differential growing neural gas · Representation learning · Modeling of neuron groups

1 Introduction

In a recent paper [12] we outlined a functional model of cortical columns [1, 18, 19] that utilized a *recursive growing neural gas* (RGNG) as one of its core components. More specifically, two reciprocally coupled RGNGs were used to describe two groups of neurons within a single cortical column that together form a local, autoassociative memory cell (AMC). Originally, we introduced the RGNG algorithm [6, 8] to model

J. Kerdels (✉) · G. Peters
FernUniversität in Hagen, University of Hagen, Human-Computer Interaction,
Faculty of Mathematics and Computer Science, Universitätsstrasse 1, 58097 Hagen, Germany
e-mail: jochen.kerdels@fernuni-hagen.de

G. Peters
e-mail: gabriele.peters@fernuni-hagen.de

© The Author(s), under exclusive license to Springer Nature Switzerland AG 2021   109
C. Sabourin et al. (eds.), *Computational Intelligence*, Studies in Computational
Intelligence 893, https://doi.org/10.1007/978-3-030-64731-5_6

the behavior of entorhinal grid cells [3, 4]. In this earlier model a single RGNG was used to describe a single group of neurons, effectively modeling one half of the later proposed cortical AMC.

Here we review the RGNG algorithm and introduce an algorithmic alternative that is structurally less complex, more efficient to compute, and more robust in terms of the input space representation that is learned by the algorithm. Given that our future research on cortical column models aims at simulating networks of hundreds to thousands of cortical columns it seems prudent to improve and optimize the model used so far in light of this new field of application, i.e., when moving away from simulating single groups of neurons towards simulating entire networks of neuron groups.

The next two sections provides a recapitulation of the RGNG algorithm highlighting its core characteristics and its usage in the context of modeling the response behavior of a group of neurons. Section 3 introduces a novel algorithm, the *differential growing neural gas*, that addresses some of the shortcomings of the RGNG while maintaining its main properties in the context of modeling neuron groups. In Sect. 4 the behavior of the introduced algorithm is demonstrated and analyzed using the well known MNIST dataset [16]. Finally, Sect. 5 concludes the paper and outlines aspects of future research.

2   RGNG Revisited

The recursive growing neural gas (RGNG) is an unsupervised learning algorithm that learns a prototype-based representation of a given input space. Although the RGNG is algorithmically similar to well known prototype-based methods of unsupervised learning like the original growing neural gas (GNG) [2, 17] or self organizing maps [15], the resulting input space representation is significantly different from common prototype-based approaches. While the latter use single prototype vectors to represent local regions of input space that are pairwise disjoint, the RGNG uses a sparse distributed representation where each point in input space is encoded by a joint ensemble activity.

From a neurobiological perspective an RGNG can be interpreted as describing the behavior of a group of neurons that receives signals from a common input space. Each of the neurons in this group tries to learn a coarse representation of the *entire* input space while being in competition with one another. This coarse representation consists of a limited number of prototypical input patterns that are learned competitively and stored on separate branches of the neuron's dendritic tree. The competitive character of the learning process ensures that the prototypical input patterns are distributed across the entire input space and thus form a coarse, pointwise representation of it. In addition to these intra-neuronal processes the inter-neuronal competition on the neuron group level influences the alignment of the individual neuron's representations. More specifically, the competition between the neurons favors an alignment of the individual representations in such a way that the different representations become pairwise distinct. As a result, the neuron group as a whole forms a dense representation of the input space consisting of the self-similar, coarse representations of its group members. The activity of individual neurons in such a group in response to a given input is ambiguous as it cannot be determined from the outside which of the learned input patterns triggered the neuron's response. However, the collective activity of all neurons in response to a shared input creates an activity pattern that is highly specific for the given input since it is unlikely that for two different inputs the response of all neurons would be exactly the same.

A formal description of the RGNG algorithm is given in the next section. It was adapted from [9]. The given description is independent of the RGNG's application in the aforementioned neurobiological context, which is described later in Sect. 2.2.

2.1 Formal Description

An RGNG g can be described by a tuple[1]:

$$g := (U, C, \theta) \in G,$$

with a set U of units, a set C of edges, and a set θ of parameters.

Each unit u is described by a tuple:

$$u := (w, e) \in U, \quad w \in W := \mathbb{R}^n \cup G, \quad e \in \mathbb{R},$$

with the *prototype* w, and the *accumulated error* e. Note that the prototype w of an RGNG unit can either be a n-dimensional vector or another RGNG.

Each edge c is described by a tuple:

$$c := (V, t) \in C, \quad V \subseteq U \land |V| = 2, \quad t \in \mathbb{N},$$

with the units $v \in V$ connected by the edge and the *age* t of the edge. The *direct neighborhood* E_u of a unit $u \in U$ is defined as:

$$E_u := \{k | \exists (V, t) \in C, \quad V = \{u, k\}, \quad t \in \mathbb{N}\}.$$

The set θ of parameters consists of:

$$\theta := \{\epsilon_b, \epsilon_n, \epsilon_r, \lambda, \tau, \alpha, \beta, M\}.$$

The behavior of an RGNG is defined by four functions. The distance function

$$D(x, y) : W \times W \to \mathbb{R}$$

---

[1]The notation $g.\alpha$ is used to reference the element α within the tuple.

determines the distance either between two vectors, two RGNGs, or a vector and an RGNG. The interpolation function

$$I(x, y) : (\mathbb{R}^n \times \mathbb{R}^n) \cup (G \times G) \rightarrow W$$

generates a new vector or new RGNG by interpolating between two vectors or two RGNGs, respectively. The adaptation function

$$A(x, \xi, r) : W \times \mathbb{R}^n \times \mathbb{R} \rightarrow W$$

adapts either a vector or RGNG towards the input vector ξ by a given fraction r. Finally, the input function

$$F(g, \xi) : G \times \mathbb{R}^n \rightarrow G \times \mathbb{R}$$

feeds an input vector ξ into the RGNG g and returns the modified RGNG as well as the distance between ξ and the best matching unit s_1 (BMU, see below) of g. The input function F contains the core of the RGNG's behavior and utilizes the other three functions, but is also used, in turn, by those functions introducing several recursive paths to the program flow.

$F(g, \xi)$. The input function F is a generalized version of the original GNG algorithm that facilitates the use of prototypes other than vectors. In particular, it allows to use RGNGs themselves as prototypes resulting in a recursive structure. An input $\xi \in \mathbb{R}^n$ to the RGNG g is processed by the input function F as follows:

– Find the two units s_1 and s_2 with the smallest distance to the input ξ according to the distance function D:

$$s_1 := \arg\min_{u \in g.U} D(u.w, \xi),$$
$$s_2 := \arg\min_{u \in g.U \setminus \{s_1\}} D(u.w, \xi).$$

– Increment the age of all edges connected to s_1:

$$\Delta c.t = 1, \quad c \in g.C \land s_1 \in c.V.$$

– If no edge between s_1 and s_2 exists, create one:

$$g.C \Leftarrow g.C \cup \{(\{s_1, s_2\}, 0)\}.$$

– Reset the age of the edge between s_1 and s_2 to zero:

$$c.t \Leftarrow 0, \quad c \in g.C \land s_1, s_2 \in c.V.$$

– Add the squared distance between ξ and the prototype of s_1 to the accumulated error of s_1:

$$\Delta s_1.e = D(s_1.w, \xi)^2.$$

- Adapt the prototype of s_1 and all prototypes of its direct neighbors:

$$s_1.w \Leftarrow A(s_1.w, \xi, g.\theta.\epsilon_b),$$
$$s_n.w \Leftarrow A(s_n.w, \xi, g.\theta.\epsilon_n), \forall s_n \in E_{s_1}.$$

- Remove all edges with an age above a given threshold τ and remove all units that no longer have any edges connected to them:

$$g.C \Leftarrow g.C \setminus \{c|c \in g.C \wedge c.t > g.\theta.\tau\},$$
$$g.U \Leftarrow g.U \setminus \{u|u \in g.U \wedge E_u = \emptyset\}.$$

- If an integer-multiple of $g.\theta.\lambda$ inputs was presented to the RGNG g and $|g.U| < g.\theta.M$, add a new unit u. The new unit is inserted "between" the unit j with the largest accumulated error and the unit k with the largest accumulated error among the direct neighbors of j. Thus, the prototype $u.w$ of the new unit is initialized as:

$$u.w := I(j.w, k.w), \quad j = \arg\max_{l \in g.U} (l.e),$$
$$k = \arg\max_{l \in E_j} (l.e).$$

The existing edge between units j and k is removed and edges between units j and u as well as units u and k are added:

$$g.C \Leftarrow g.C \setminus \{c|c \in g.C \wedge j, k \in c.V\},$$
$$g.C \Leftarrow g.C \cup \{(\{j, u\}, 0), (\{u, k\}, 0)\}.$$

The accumulated errors of units j and k are decreased and the accumulated error $u.e$ of the new unit is set to the decreased accumulated error of unit j:

$$\Delta j.e = -g.\theta.\alpha \cdot j.e, \quad \Delta k.e = -g.\theta.\alpha \cdot k.e,$$
$$u.e := j.e.$$

- Finally, decrease the accumulated error of all units:

$$\Delta u.e = -g.\theta.\beta \cdot u.e, \quad \forall u \in g.U.$$

The function F returns the tuple $(g, d_{\min})$ containing the now updated RGNG g and the distance $d_{\min} := D(s_1.w, \xi)$ between the prototype of unit s_1 and input ξ. Note that in contrast to the regular GNG there is no stopping criterion any more, i.e., the RGNG operates explicitly in an online fashion by continuously integrating new inputs. To prevent unbounded growth of the RGNG the maximum number of units $\theta.M$ was introduced to the set of parameters.

   $D(x, y)$. The distance function D determines the distance between two prototypes x and y. The calculation of the actual distance depends on whether x and y are both vectors, a combination of vector and RGNG, or both RGNGs:

$$D(x, y) := \begin{cases} D_{RR}(x, y) & \text{if } x, y \in \mathbb{R}^n, \\ D_{GR}(x, y) & \text{if } x \in G \wedge y \in \mathbb{R}^n, \\ D_{RG}(x, y) & \text{if } x \in \mathbb{R}^n \wedge y \in G, \\ D_{GG}(x, y) & \text{if } x, y \in G. \end{cases}$$

In case the arguments of D are both vectors, the Minkowski distance is used:

$$D_{RR}(x, y) := \left(\sum_{i=1}^n |x_i - y_i|^p \right)^{\frac{1}{p}}, \quad \begin{aligned} x &= (x_1, \ldots, x_n), \\ y &= (y_1, \ldots, y_n), \\ p &\in \mathbb{N}. \end{aligned}$$

Using the Minkowski distance instead of the Euclidean distance allows to adjust the distance measure with respect to certain types of inputs via the parameter p. For example, setting p to higher values results in an emphasis of large changes in individual dimensions of the input vector versus changes that are distributed over many dimensions [7]. However, in the common case the parameter is set to a fixed value of 2 which makes the Minkowski distance equivalent to the Euclidean distance.

In case the arguments of D are a combination of vector and RGNG, the vector is fed into the RGNG using function F and the returned minimum distance is taken as distance value:

$$D_{GR}(x, y) := F(x, y).d_{\min},$$
$$D_{RG}(x, y) := D_{GR}(y, x).$$

In case the arguments of D are both RGNGs, the distance is defined to be the pairwise minimum distance between the prototypes of the RGNGs' units, i.e., *single linkage* distance between the sets of units is used:

$$D_{GG}(x, y) := \min_{u \in x.U, k \in y.U} D(u.w, k.w).$$

The latter case is used by the interpolation function if the recursive depth of an RGNG is at least 2. In a neurobiological context (see 2.2) an RGNG-based model typically has only a recursive depth of 1. Hence, the case is considered for reasons of completeness rather than necessity. Alternative measures to consider could be, e.g.., *average* or *complete* linkage.

$I(x, y)$. The interpolation function I returns a new prototype as a result from interpolating between the prototypes x and y. The type of interpolation depends on whether the arguments are both vectors or both RGNGs:

$$I(x, y) := \begin{cases} I_{RR}(x, y) & \text{if } x, y \in \mathbb{R}^n, \\ I_{GG}(x, y) & \text{if } x, y \in G. \end{cases}$$

In case the arguments of I are both vectors, the resulting prototype is the arithmetic mean of the arguments:

$$I_{RR}(x, y) := \frac{x + y}{2}.$$

In case the arguments of I are both RGNGs, the resulting prototype is a new RGNG a. Assuming w.l.o.g. that $|x.U| \geq |y.U|$ the components of the interpolated RGNG a are defined as follows:

$$a := I(x, y),$$

$$a.U := \left\{ (w, 0) \;\middle|\; \begin{array}{l} w = I(u.w, k.w), \\ \forall u \in x.U, \\ k = \arg\min_{l \in y.U} D(u.w, l.w) \end{array} \right\},$$

$$a.C := \left\{ (\{l, m\}, 0) \;\middle|\; \begin{array}{l} \exists c \in x.C \\ \wedge \;\; u, k \in c.V \\ \wedge \;\; l.w = I(u.w, \cdot) \\ \wedge \;\; m.w = I(k.w, \cdot) \end{array} \right\},$$

$$a.\theta := x.\theta.$$

The resulting RGNG a has the same number of units as RGNG x. Each unit of a has a prototype that was interpolated between the prototype of the corresponding unit in x and the nearest prototype found in the units of y. The edges and parameters of a correspond to the edges and parameters of x.

$A(x, \xi, r)$. The adaptation function A adapts a prototype x towards a vector ξ by a given fraction r. The type of adaptation depends on whether the given prototype is a vector or an RGNG:

$$A(x, \xi, r) := \begin{cases} A_R(x, \xi, r) & \text{if } x \in \mathbb{R}^n, \\ A_G(x, \xi, r) & \text{if } x \in G. \end{cases}$$

In case prototype x is a vector, the adaptation is performed as linear interpolation:

$$A_R(x, \xi, r) := (1 - r)x + r\xi.$$

In case prototype x is an RGNG, the adaptation is performed by feeding ξ into the RGNG. Importantly, the parameters ϵ_b and ϵ_n of the RGNG are temporarily changed to take the fraction r into account:

$$\theta^* := (\, r, \;\; r \cdot x.\theta.\epsilon_r, \;\; x.\theta.\epsilon_r, \;\; x.\theta.\lambda, \;\; x.\theta.\tau,$$
$$x.\theta.\alpha, \;\; x.\theta.\beta, \;\; x.\theta.M)\,,$$
$$x^* := (x.U, \; x.C, \; \theta^*)\,,$$
$$A_G(x, \xi, r) := F(x^*, \xi).x\,.$$

Note that in this case the new parameter $\theta.\epsilon_r$ is used to derive a temporary ϵ_n from the fraction r.

This concludes the formal definition of the RGNG algorithm.

2.2 RGNG-based Neuron Model

The RGNG algorithm as described above is used in the context of modeling a group of neurons as follows: The recursive depth of the RGNG is limited to one, which results in a two-layered structure with layers $L1$ and $L2$. The RGNG units of layer $L1$ correspond to the individual neurons of the modeled group. Each RGNG unit in $L1$ has a prototype that is a separate RGNG located in layer $L2$. This separate RGNG represents the dendritic tree of the corresponding neuron.

The RGNG in layer $L1$ is parameterized by θ_1. All RGNGs in layer $L2$ are parameterized by θ_2. Hence, the number of units (*neurons*) in $L1$ is set by $\theta_1.M$. The number of units (*prototypical input patterns in one dendritic tree*) per $L1$ unit is set by $\theta_2.M$. Therefore, the input space representation learned by the group of neurons as a whole utilizes $\theta_1.M \times \theta_2.M$ prototypical input patterns located in layer $L2$.

Inputs $\xi \in \mathbb{R}^n$ to the neuron model are processed through the input function F of the RGNG in $L1$. The resulting call graph is depicted in Fig. 1. Given that functions D and A are linear in n, and function I is constant amortized the computational cost to process a single, n-dimensional input by a two layer RGNG is $O\ (n \cdot \theta_1.M \cdot \theta_2.M)$.

After an input ξ is processed the output of the modeled group of $K := \theta_1.M$ neurons is given as an ensemble activity $\mathbf{a} := (a_0, \ldots, a_{K-1})$ using a softmax function:

$$a_i := \frac{e^{\hat{a}_i}}{\sum_{j=0}^{K-1} e^{\hat{a}_j}},$$

with

$$\hat{a}_i := \gamma \left(1 - \frac{\|\mathbf{s}_1^i - \xi\|_2}{\|\mathbf{s}_2^i - \xi\|_2} \right), \quad i = 0, \ldots, K-1,$$

and $\mathbf{s}_1^i, \mathbf{s}_2^i$ being the best and second best matching prototypical input patterns found in the dendritic tree of neuron i, which are identified during the computation of function F of the corresponding $L2$ RGNG. The factor γ is used to control the degree with which the softmax function emphasizes the largest elements of the ensemble activity $\mathbf{a}$. From a neurobiological perspective the parameter γ can be interpreted as the strength of local inhibition that the neurons with the highest activations exert on the other neurons of the group.

The RGNG-based neuron group model was used successfully to describe the response behavior of typical entorhinal grid cells [6] as well as entorhinal cells that show grid like activity in response to saccadic eye movements [9, 14]. Furthermore, the model was analyzed regarding its resilience to noise [10, 13], and its ability to perform pattern separation [11]. Lastly, it was integrated into a model of cortical column function as part of an autoassociative memory cell [12].

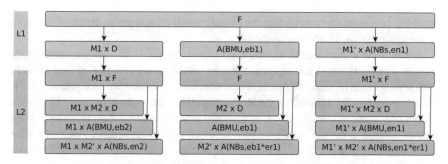

Fig. 1 Call graph for a single call to the input function F of an RGNG with two layers ($L1$, $L2$). Executing F in layer $L1$ results in $\theta_1 \cdot M$ (M1) calls to function D, a single call to function A with learning rate $\theta_1 \cdot \epsilon_b$ (eb1), and $O\,(\theta_1 \cdot M - 1)$ (M1') calls to function A with learning rate $\theta_1 \cdot \epsilon_n$ (en1), where $O\,(\theta_1 \cdot M - 1)$ is the potential size of the direct neighborhood of the corresponding BMU. The calls to D and A result in recursive calls to F on the next lower layer $L2$. Note that functions A temporarily change the learning rates for their calls to F. The recursion stops when a layer is reached where the prototypes are vectors; here $L2$. Calls to function I while the RGNGs are still growing are not shown since their computational cost is constant amortized. Figure adapted from [6]

The aforementioned use cases of the RGNG-based neuron model focused on describing the response behavior of single neuron groups. Building on our latest work in modeling a cortical column [12] we aim to model large networks of cortical columns in our future research, i.e., to model hundreds or thousands of neuron groups. To this end it seems prudent to attempt to optimize the RGNG-based model with respect to its complexity, robustness, and computational requirements.

3 RGNG Approximation

The way the RGNG-based neuron group model forms a distributed, prototype-based representation of the group's input space is the key aspect of this neuron model. Two main processes are involved in the formation of this representation. First, each neuron has to learn a representation of the entire input space, i.e., it has to distribute the prototypical input patterns it learns across the input space rather than specialize in any one local region. Second, the representations of individual neurons have to be aligned in such a way that they are pairwise distinct and enable a disambiguation of inputs via the ensemble activity of the neuron group. In the RGNG-based model these two processes depend both on the dynamics encoded in input function F and are controlled by the learning rates $\theta_1 \cdot \epsilon_b / \cdot \epsilon_n$ and $\theta_2 \cdot \epsilon_b / \cdot \epsilon_n$ on layers $L1$ (group level alignment) and $L2$ (per neuron learning), respectively. Although the underlying growing neural gas algorithm is relatively robust regarding the choice of learning rates, it can still be difficult to identify a set of learning rates that ensure both the distribution of per-neuron prototypes across the input space and the alignment of the resulting representations such that the overall group-level representation of the input space settles into a stable configuration.

Another important aspect of the RGNG-based model is its computational complexity. For every input the algorithm has to calculate the distance of that input to each prototypical input pattern of every neuron. Since the distance calculation itself is not computationally demanding the RGNG algorithm becomes effectively I/O bound on typical, modern computer systems. When scaling a simulation from individual neuron groups to networks of hundreds or thousands of groups, i.e., exceeding the cache capacity of the system in use the I/O boundedness of the algorithm becomes the main limiting factor of that simulation regarding computation time.

To address both of these issues we present a novel variation of the growing neural gas algorithm, the *differential growing neural gas* (DGNG), and propose to use it as an approximation of the RGNG algorithm in the context of neuron group modeling. The main idea of the DGNG algorithm is to partition the input space into N regions, where N corresponds to the number of prototypical input patterns that are learned by each neuron. This partition is learned by a top layer growing neural gas with N units representing each input space region by a single prototype vector located at the center of that region. Each region is then partitioned by separate sub-DGNGs into K sub-regions, where K corresponds to the number of neurons in the modeled neuron group. The units of these sub-DGNGs contain prototype vectors that represent a position relative to the center prototype of the corresponding region, i.e., the prototype vectors encode the respective difference between input and center prototype for a given input space sub-region.

Compared to the RGNG-based model the correspondence between $L1$ units and the neurons of the modeled neuron group has been removed in the DGNG. Instead, the representation of each neuron is now distributed among the different sub-DGNGs of the top layer DGNG units. More precisely, the i-th unit of sub-DGNG j contains the j-th prototype of neuron i.

Whereas the distribution and alignment of prototypes in the RGNG algorithm depend on a suitable choice of four different learning rates, the distribution and alignment of prototypes in the DGNG algorithm is directly enforced by its structure, thereby reducing the dependence on suitable learning rates for a particular problem instance. The partitioning of the input space in layer $L1$ ensures that the coarse input space representations of all neurons always cover the entire input space, while the competition in the $L2$ sub-DGNGs ensures that the representations learned by each neuron are pairwise distinct. In addition, the partitioning of the input space allows to reduce the number of distance calculations per input significantly (see Sect. 3.2).

Analogous to Sect. 2, the next section will introduce the DGNG algorithm independent of its use for modeling a group of neurons. Subsequently, Sect. 3.2 will describe in more detail how the DGNG is used to describe the response behavior of a group of neurons.

3.1 DGNG Formal Description

Like an RGNG a DGNG g can be described by a tuple:

$$g := (U, C, \theta) \in G,$$

with a set U of units, a set C of edges, and a set θ of parameters. Each unit u is described by a tuple:

$$u := (w, e, s) \in U, \quad w \in \mathbb{R}^n, \quad e \in \mathbb{R}, \quad s \in G \cup \emptyset,$$

with the *prototype* w, the *accumulated error* e, and the *sub-DGNG* s. Note that in contrast to the RGNG the prototype w of an DGNG unit is always a n-dimensional vector, and the recursive structure is explicitly given by the sub-DGNG s, which is empty at the lowest level.

Each edge c is described by a tuple:

$$c := (V, t) \in C, \quad V \subseteq U \wedge |V| = 2, \quad t \in \mathbb{N},$$

with the units $v \in V$ connected by the edge and the *age* t of the edge. The *direct neighborhood* E_u of a unit $u \in U$ is defined as:

$$E_u := \{k | \exists (V, t) \in C, \quad V = \{u, k\}, \quad t \in \mathbb{N}\}.$$

The set θ of parameters consists of:

$$\theta := \{\epsilon_b, \epsilon_n, \lambda, \tau, \alpha, \beta, M\}.$$

In contrast to the RGNG the behavior of a DGNG can be described by a single (recursive) input function F that feeds an input vector ξ into the DGNG g and returns the modified DGNG as well as the prototype vector w^* of the best matching unit:

$$F(g, \xi) : G \times \mathbb{R}^n \to G \times \mathbb{R}^n.$$

Note the subtle difference that this input function does not return the minimum distance to the input but rather the best matching prototype vector.

$F(g, \xi)$. The input function F processes an input $\xi \in \mathbb{R}^n$ to the DGNG g as follows:

- If $g = \varnothing$ return g and a zero vector $\mathbf{0}$ as best matching prototype, else:
- Find the two units s_1 and s_2 with the smallest distance to the input ξ:

$$s_1 := \arg\min_{u \in g.U} \|u.w - \xi\|_p,$$
$$s_2 := \arg\min_{u \in g.U \setminus \{s_1\}} \|u.w - \xi\|_p.$$

- Increment the age of all edges connected to s_1:

$$\Delta c.t = 1, \quad c \in g.C \wedge s_1 \in c.V .$$

- If no edge between s_1 and s_2 exists, create one:

$$g.C \Leftarrow g.C \cup \{(\{s_1, s_2\}, 0)\} .$$

- Reset the age of the edge between s_1 and s_2 to zero:

$$c.t \Leftarrow 0, \quad c \in g.C \wedge s_1, s_2 \in c.V .$$

- Add the squared distance between ξ and the prototype of s_1 to the accumulated error of s_1:

$$\Delta s_1.e = \|s_1.w - \xi\|_p^2.$$

- Adapt the prototype of s_1 and all prototypes of its direct neighbors:

$$\Delta s_1.w = (\xi - (s_1.w + F(s_1.s, \xi - s_1.w).w^*)) \, g.\theta.\epsilon_b,$$
$$\Delta s_n.w = (\xi - s_n.w) \, g.\theta.\epsilon_n.$$

Note that in this step a recursion takes place. The sub-DGNG s of the best matching unit s_1 is fed with the difference between the input ξ and the prototype w of s_1, and the returned best matching prototype w^* of the sub-DGNG s is used to augment $s_1.w$ in the current adaptation step. Thus, the sub-DGNG s can be understood as learning the details or different variations of the more coarse representation stored in $s_1.w$. The adaptation of the neighboring units is performed without recursion to reduce computational cost.

- Next, remove all edges with an age above a given threshold τ and remove all units that no longer have any edges connected to them:

$$g.C \Leftarrow g.C \setminus \{c \mid c \in g.C \wedge c.t > g.\theta.\tau\},$$
$$g.U \Leftarrow g.U \setminus \{u \mid u \in g.U \wedge E_u = \emptyset\}.$$

- If an integer-multiple of $g.\theta.\lambda$ inputs was presented to the RGNG g and $|g.U| < g.\theta.M$, add a new unit u. The new unit is inserted "between" the unit j with the largest accumulated error and the unit k with the largest accumulated error among the direct neighbors of j. Thus, the prototype $u.w$ of the new unit is initialized as:

$$u.w := \frac{j.w + k.w}{2}, \quad j = \arg\max_{l \in g.U} (l.e),$$
$$k = \arg\max_{l \in E_j} (l.e).$$

The existing edge between units j and k is removed and edges between units j and u as well as units u and k are added:

$$g.C \Leftarrow g.C \setminus \{c | c \in g.C \wedge j, k \in c.V\},$$
$$g.C \Leftarrow g.C \cup \{((\{j, u\}, 0), (\{u, k\}, 0)\}.$$

The accumulated errors of units j and k are decreased and the accumulated error $u.e$ of the new unit is set to the decreased accumulated error of unit j:

$$\Delta j.e = -g.\theta.\alpha \cdot j.e, \quad \Delta k.e = -g.\theta.\alpha \cdot k.e,$$
$$u.e := j.e.$$

- Finally, decrease the accumulated error of all units:

$$\Delta u.e = -g.\theta.\beta \cdot u.e, \quad \forall u \in g.U.$$

The function F returns the tuple (g, w^*) containing the now updated DGNG g and the prototype w^* of the best matching unit s_1 w.r.t. input ξ.

This concludes the formal definition of the DGNG algorithm.

3.2 DGNG-based Neuron Model

When modeling a group of neurons with the DGNG algorithm the recursive depth of the DGNG is limited to one like it is the case when using an RGNG. However, the resulting two-layered structure is interpreted differently. There exists no longer a direct correspondence between certain DGNG units and neurons of the modeled neuron group. Instead, the prototypical input patterns learned by an individual neuron are distributed across the sub-DGNGs of all DGNG units of layer $L1$ and are composed as summation of the respective $L1$ unit's prototype vector and one prototype vector of the sub-DGNG's $L2$ units. Hence, the number of DGNG units in $L1$ corresponds to the number of prototypical input patterns learned by one neuron in it's dendritic tree, and the number of DGNG units in each sub-DGNG, i.e., in layer $L2$ corresponds to the number of neurons in the neuron group.

More specifically, given a group of K neurons that each learn N prototypical input patterns the set of prototypes $P^i := \{p_0^i, \ldots, p_{N-1}^i\}$ of neuron i within a DGNG g is defined as:

$$p_j^i := u_j.w + u_j.s.v_i, \quad u_j \in g.U, \quad v_i \in u_j.s.U, \quad j = 0, \ldots, N-1,$$

with $g.\theta.M = N$ and $u.s.\theta.M = K$, $\forall u \in g.U$. The computational complexity of processing an input $\xi \in \mathbb{R}^n$ with the DGNG-based model is significantly reduced compared to using an RGNG. Instead of requiring $O(n \cdot K \cdot N)$ operations the DGNG-based solution requires only $O(n(K+N))$ operations.

The ensemble activity $\mathbf{a} := (a_0, \ldots, a_{K-1})$ of a DGNG-based neuron model is based on the sub-DGNGs of the best and second best matching DGNG units s_1 and s_2 in $L1$ and uses, like the RGNG-based model, a softmax function:

$$a_i := \frac{e^{\hat{a}_i}}{\sum_{j=0}^{K-1} e^{\hat{a}_j}},$$

with

$$\hat{a}_i := \gamma \left(1 - \frac{\|s_1.s.u_i.w - (\xi - s_1.w)\|_2}{\|s_2.s.u_i.w - (\xi - s_2.w)\|_2} \right), \quad i = 0, \ldots, K-1.$$

As with the RGNG-based model the factor γ is used to control the degree with which the softmax function emphasizes the largest elements of the ensemble activity $\mathbf{a}$.

4   Results

In order to perform a first characterization of a DGNG-based neuron group model we set up a model with 100 neurons, each of which had the capacity of learning 16 prototypical input patterns. We trained the model with inputs from the well-known MNIST dataset [16], which consists of 60000 grayscale images of handwritten digits with a resolution of 28×28 pixels. As parameters we used: $\theta.\epsilon_b := 0.01$, $\theta.\epsilon_b := 0.0001$, $\theta.\lambda := 1000$, $\theta.\tau := 300$, $\theta.\alpha := 0.5$, $\theta.\beta := 0.0005$, $\theta.M := \{16, 100\}$. We presented the MNIST training dataset repeatedly to the model until 10 million inputs were reached in total. After training the response of the modeled neuron group to the MNIST test dataset (10000 handwritten digits that were not in the training dataset) was stored as a set of ensemble activity vectors that were used in the following analysis. The ensemble activity vectors were sampled multiple times for varying values of output parameter γ (Sect. 3.2).

The 16 prototypical input patterns learned by the DGNG in $L1$ that partition the input space are shown in Fig. 2a. As expected, the prototypes average over large regions of input space and thus show only vague patterns of typical, handwritten digits. Figure 2b shows the *relative* prototypes of a single modeled neuron. Each of these prototypes were learned by a different $L2$ DGNG associated with the corresponding

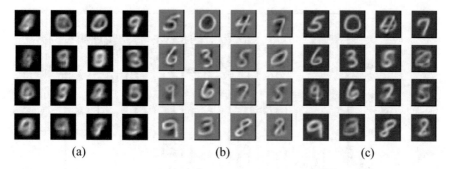

Fig. 2 Examples of learned prototypical input patterns. (**a**) Prototypes learned by the DGNG in $L1$ that partition the input space. (**b**) Relative prototypes of a single modeled neuron learned by different $L2$ DGNGs in the respective $L1$ input space regions. (**c**) Summation of the prototypes shown in **a** and **b** resulting in the actual patterns that are compared with the particular inputs

$L1$ input space region and prototype. These relative $L2$ prototypes learn the difference between the average $L1$ prototype of their associated input space region and the more local input space region they represent. Combined, the average $L1$ prototype and the relative $L2$ prototype result in the patterns shown in Fig. 2c, which are those that are effectively used to determine the particular distance to a given input pattern and derive the activity of the particular neuron in response to that input. Figure 2c also illustrates that neurons in this neuron model do not specialize in a single type of input pattern, e.g.., the digit 0, but respond to a variety of different input patterns such that a disambiguation of different input patterns has to happen at the neuron group level on the basis of the group's ensemble activity.

In the outlined DGNG-based neuron group model the group's ensemble activity in response to an input is a vector $\mathbf{a} \in \mathbb{R}^K$ with K being the number of neurons in the modeled group. The output parameter γ (Sect. 3.2) controls the sparsity of this activity vector $\mathbf{a}$, which in turn determines the specificity of the neuron group's response to a given input. To analyze the effect the parameter γ has on the sparsity of $\mathbf{a}$ we calculated the distribution of Gini Indices of the ensemble activity vectors that were returned in response to the MNIST test dataset. The resulting distribution as a function of γ is shown in Fig. 3. The Gini Index can be used as a measure for the sparsity of a vector [5]. As an approximation, the value of the Gini Index can be intuitively understood as the fraction of entries in a given vector that have low values (compared to the other entries), i.e., a Gini Index of 0 corresponds to a vector that has similar values in all of its entries whereas a Gini Index of 1 corresponds to a vector that has one or only a few entries with values significantly higher than all other entries of the vector.

The distribution shown in Fig. 3 illustrates that with increasing values of γ the mean sparseness of the ensemble activity vectors steadily increases as well. Considering the interpretation of γ as the degree of local inhibition in the proposed neuron model it becomes evident how important this local inhibition is to the formation of a sparse ensemble code that represents a given input with a sufficient degree of

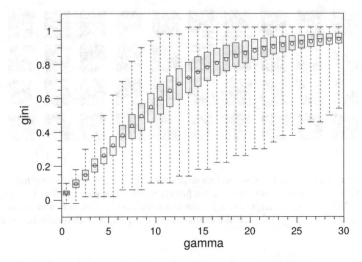

Fig. 3 Distribution of Gini Indices of the ensemble activity vectors of the modeled neuron group in response to the MNIST test dataset as a function of output parameter *gamma* (Sect. 3.2). Whiskers indicate upper and lower quartiles, circles and center horizontal lines indicate mean and median of the distributions

specificity. The ability to perform such a pattern separation was already a key characteristic of the RGNG-based neuron model [11]. To investigate the DGNG-based model in that regard we compared the pairwise cosine similarities of the ensemble activity vectors that were generated in response to the MNIST test dataset. Since the dataset provides labels for the ten different digit classes it was possible to compare the cosine similarities for intra- and inter-class inputs separately. Given that intra-class samples are likely to be more similar to each other than inter-class samples this distinction of cases allows to study the pattern separation abilities in more detail. In general it is expected that the group of neurons is able to represent individual inputs distinctively in both cases, although the task in the intra-class case is conceivably harder.

Figure 4 shows the resulting distributions of cosine similarities. In both cases the ensemble activity vectors for values of γ below 5 are very similar to each other and do not allow to distinguish the neuron group's responses to different inputs very well—again emphasizing the importance of local inhibition in this neuron model. However, with increasing values of γ (≥ 5), i.e., increasing local inhibition, the ensemble activity rapidly becomes very specific. Interestingly, the increase in specificity is not monotonic. The distributions for both intra-class and inter-class samples exhibit a minimum at $\gamma \approx 7$ and $\gamma \approx 5$ to 15, respectively. For these values of γ the response of the neuron group to a given input ξ is such that very few to none of the other test inputs have a cosine similarity close to one with respect to ξ while the spectrum of occurring, lower cosine similarity values is wide. In contrast, for values of $\gamma > 15$ the distinction between similar or dissimilar inputs becomes much more pronounced. The vast majority of other test inputs (note the logarithmic scale) exhibit cosine similarity

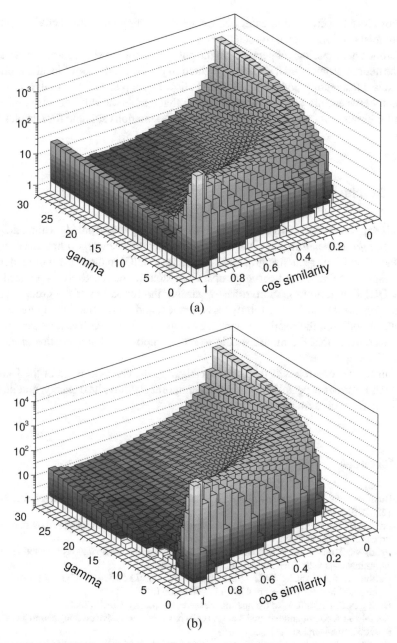

Fig. 4 Histograms of pairwise cosine similarities between ensemble activity vectors of the modeled neuron group in response to the MNIST test dataset. Figure (**a**) shows cosine similarities for intra-class samples, while figure (**b**) shows the results for inter-class samples. In both cases, increasing values of γ lead to stronger pattern separation. Note the logarithmic scale of the z-axis

values close to zero in that case, i.e., the corresponding ensemble activity vectors become almost orthogonal to each other.

From a neurobiological perspective the putative ability to shape the characteristic of the neuron group's input space representation just by the degree of local inhibition alone is a fascinating possibility. It would allow a group of neurons to dynamically switch between fine grained representations that enable the differentiation of very similar inputs and coarse grained, clear-cut representations suitable for fast classification.

5   Conclusion

In this paper we presented a novel variant of the growing neural gas algorithm: the *differential growing neural gas* (DGNG). We designed this new algorithm as a drop-in replacement of the recursive growing neural gas (RGNG) in the context of modeling the response behavior of neuron groups. Compared to the old RGNG approach the new DGNG algorithm is more robust regarding the formation of the group's input space representation, is structurally less complex, and is computationally more efficient. In addition, the results of our first analysis of a DGNG-based neuron group model indicate that the model is able to retain important characteristics of earlier RGNG-based models.

Our future research will focus on investigating the characteristics of the DGNG algorithm further especially regarding its temporal stability, its capacity, and its use in our cortical column model.

References

1. Buxhoeveden, D.P., Casanova, M.F.: The minicolumn hypothesis in neuroscience. Brain **125**(5), 935–951 (2002)
2. Fritzke, B.: A growing neural gas network learns topologies. In: Advances in Neural Information Processing Systems 7. pp. 625–632. MIT Press (1995)
3. Fyhn, M., Molden, S., Witter, M.P., Moser, E.I., Moser, M.B.: Spatial representation in the entorhinal cortex. Science **305**(5688), 1258–1264 (2004)
4. Hafting, T., Fyhn, M., Molden, S., Moser, M.B., Moser, E.I.: Microstructure of a spatial map in the entorhinal cortex. Nature **436**(7052), 801–806 (2005)
5. Hurley, N.P., Rickard, S.T.: Comparing measures of sparsity. CoRR (2008)
6. Kerdels, J.: A computational model of grid cells based on a recursive growing neural gas. Ph.D. thesis, FernUniversität in Hagen (2016)
7. Kerdels, J., Peters, G.: Analysis of high-dimensional data using local input space histograms. Neurocomputing **169**, 272–280 (2015)
8. Kerdels, J., Peters, G.: A new view on grid cells beyond the cognitive map hypothesis. In: 8th Conference on Artificial General Intelligence (AGI 2015) (July 2015)
9. Kerdels, J., Peters, G.: Modelling the grid-like encoding of visual space in primates. In: Proceedings of the 8th International Joint Conference on Computational Intelligence, IJCCI 2016, Volume 3: NCTA, Porto, Portugal, November 9–11, 2016, pp. 42–49 (2016)

10. Kerdels, J., Peters, G.: Noise resilience of an rgng-based grid cell model. In: Proceedings of the 8th International Joint Conference on Computational Intelligence, IJCCI 2016, Volume 3: NCTA, Porto, Portugal, November 9–11, 2016. pp. 33–41 (2016)
11. Kerdels, J., Peters, G.: Entorhinal grid cells may facilitate pattern separation in the hippocampus. In: Proceedings of the 9th International Joint Conference on Computational Intelligence, IJCCI 2017, Funchal, Madeira, Portugal, November 1–3, 2017. pp. 141–148 (2017)
12. Kerdels, J., Peters, G.: A grid cell inspired model of cortical column function. In: Proceedings of the 10th International Joint Conference on Computational Intelligence, pp. 204–210. INSTICC, SciTePress (2018)
13. Kerdels, J., Peters, G.: A Noise Compensation Mechanism for an RGNG-Based Grid Cell Model, pp. 263–276. Springer International Publishing (2019)
14. Kerdels, J., Peters, G.: A Possible Encoding of 3D Visual Space in Primates, pp. 277–295. Springer International Publishing (2019)
15. Kohonen, T.: Self-organized formation of topologically correct feature maps. Biol. Cybern. **43**(1), 59–69 (1982)
16. Lecun, Y., Bottou, L., Bengio, Y., Haffner, P.: Gradient-based learning applied to document recognition. Proc. IEEE **86**(11), 2278–2324 (1998)
17. Martinetz, T.M., Schulten, K.: Topology representing networks. Neural Netw. **7**, 507–522 (1994)
18. Mountcastle, V.B.: The columnar organization of the neocortex. Brain **120**(4), 701–722 (1997)
19. Mountcastle, V.B.: An organizing principle for cerebral function: The unit model and the distributed system. In: Edelman, G.M., Mountcastle, V.V. (eds.) The Mindful Brain, pp. 7–50. MIT Press, Cambridge, MA (1978)

Using Convolutional Neural Networks and Recurrent Neural Network for Human Gesture Recognition and Problem Solving

Sanghun Bang and Charles Tijus

Abstract Problem solving, such as Tower of Hanoi, is a mental process that involves identifying problems, developing strategies and organizing knowledge. We have acquired a series of data from participants solving the Tower of Hanoi problem to understand human behavior and cognitive reasoning in the problem-solving process. These data have a lot of semantic information, including the objects around participants, actions and interactions with objects. Therefore, given a series of data that are all labeled with a multiple categories, we have predicted these categories for a novel set of test data with the help of convolutional neural networks (CNNs). In addition to category inference, we use the Long Short-Term Memory(LSTM) model to do behavioral reasoning from sequential data.

Keywords Problem solving · Long Short-Term Memory · Recurrent Neural network · Convolutional Neural Network · Reinforcement learning · Cognition · Embodied cognition · Tower of Hanoi

1 Introduction

The theory of embodied cognition [2, 11] suggests that our body influences our thinking. Embodied cognition approaches made contributions to our understanding of the nature of gestures and how they influence learning. Frequently in the literature

S. Bang (✉) · C. Tijus
Laboratoire Cognitions Humaine et Artificielle (CHArt), University Paris 8, 2 rue de la Liberté, 93526 Saint-Denis, France
e-mail: sang-hun.bang@univ-paris8.fr
URL: http://www.lutin-userlab.fr

C. Tijus
e-mail: charles.tijus@univ-paris8.fr

on embodied cognition [7], gestures are used as grounding for a mapping between thinking and real objects in the world, in order for the easy catching of meanings.

Classical puzzle-like problem, such as Tower of Hanoi puzzle and missionaries-cannibals have received a lot of attentions because they do not involve domain-specific knowledge and can, therefore, be used to investigate basic cognitive mechanisms such as search and decision-making mechanisms [8]. To analyze the effect of gestures on problem solving cognitive processes (learning, memorizing, planning, and decision-making), we recruited participants who were asked to solve the puzzle of Tower of Hanoi (TOH).

Our hypothesis is that we can model the solving processes of the Tower of Hanoi, not simply through the description of the disks' moves according to the rules, but through observing the movements of the solver's hand with or without the disks. In order to test this hypothesis, we carried out an experiment for which participants were given two successive tasks: to solve the three-disk Tower of Hanoi task, then to solve this problem with four disks.

We investigated how gestures ground the meaning of abstract representations used in this experiment. The gestures added action information to their mental representation. The deictic gesture used in this experiment forces the participants to remember what they have done in previous attempt. The purpose of our research through this experiment is to infer the problem solving or the rules of game through modeling of human behavior. We collected all of the sequential gestures data that bring reaching the goal. These data were used to model and simulate how to solve the problem of TOH with Convolutional Neural Networks and Long Short-Term Memory (LSTM).

Recently, Deep CNNs [12, 14] have been very successful on single-label object classification, i.e., ImageNet Large Scale Visual Recognition Challenge [12, 13]. These algorithms have been very successful for a variety of tasks including image classification [1, 16, 17], object detection [18, 19], and others [4]. In this paper, we pay our attention to the multi-label image classification to understand the Tower of Hanoi solution. The objects around us, surrounding scenes, actions, and interactions with objects all contain semantic information as like our everyday life. The images obtained from real world have many and complex categories. Likewise, our collected sequential images also contain various semantic information: The objects around participant, theirs actions, and interactions with objects. Therefore, the task with multi-label image classification helps to understand more complex semantic information.

State-of-the-art have recently demonstrated performance' LSTM models across a variety of tasks in domains such as text [10], motion capture data [9], and music [5]. In particular, LSTM can be trained for sequence activation while processing real data sequences. Therefore, we modeled the Tower of Hanoi solving processes with the help of LSTM method.

2 Background and Related Work

2.1 Tower of Hanoi

The french mathematician Edouard Lucas introduced the Tower of Hanoi (TOH) puzzle in 1883 [3]. Figure 1 shows a standard example of TOH. There are three pegs, A, B, and C. There are four disks (d_1, d_2, d_3, d_4) on peg A. The largest disk is at the bottom of peg A and the smallest at the top. The goal of TOH is to move the whole stack of disks from the initial source peg A to a destination peg C. There are three rules as constraints: One disk at a time should be moved, in a location, the smallest disk is the one to take and a large disk cannot be placed on top of a smaller one. The minimum number of moves needed to solve TOH with n disks is denoted by $2^n - 1$.

2.2 Recurrent Neural Network

In our previous work [20], we used a simple Recurrent Neural Network which is based on Elman network [6]. As is well known, the simplest RNN model has a vanishing gradient problem. That's why, we implemented the gated activation functions, such as the long short-term Memory(LSTM) [21] to overcome the limitations of our model. Because the TOH solving process is a sequential process. In this work, we have used a LSTM-RNN network which computes a mapping from an $\mathbf{x} = (x_1, ..., x_T)$ a input sequence to an output vector sequence denoted as $\mathbf{y} = (y_1, ..., y_T)$ by calculating the network unit activation using the following equations:

Fig. 1 Four disks in Tower of Hanoi puzzle

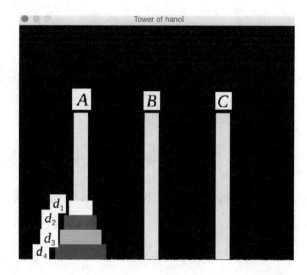

$$i_t = \delta(W_{ix}x_t + U_i h_{t-1} + b_i) \tag{1}$$

$$f_t = \delta(W_{fx}x_t + U_f h_{t-1} + b_f) \tag{2}$$

$$o_t = \delta(W_{fx}x_t + U_f h_{t-1} + b_o) \tag{3}$$

$$c_t = f_t \circ c_{t-1} + i_t \gamma(W_{cx}x_t + U_c h_{t-1} + b_c) \tag{4}$$

$$h_t = o_t \circ \delta(c_t) \tag{5}$$

where W and U terms denote weight matrices, the b terms denote bias vectors(ex: b_i is the input gate bias vector), σ and γ denote activation functions, x_t is input vector, f_t is forget gate's activation vector, i_t is input gate's activation vector, h_t is hidden state vecor, and o_t denotes output gate's activation vector.

2.3 Convolutional Neural Networks

Convolutional neural networks(CNNs) are a type of artificial neural network designed to be easy to apply to images. CNNs were first proposed in 1998 by Lecun et al [14] and consists of a convolution layer and a pooling layer, unlike the structure used in general multilayer perceptron (see Fig. 2). Convolutional Neural Networks are a very famous neural network model used of image classification. In particular, We were interested in work of Gong et. al [15] on the recent multi-label deep convolutional ranking net. Because we should take into account various situations and behaviors in the problem solving process.

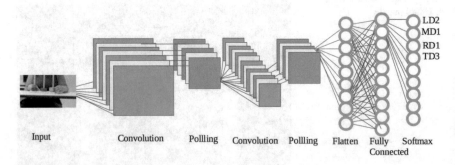

Fig. 2 The Deep Convolutional Neural Network for multi-label classification

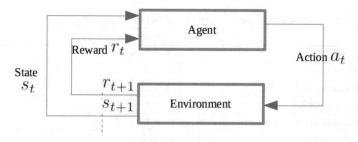

Fig. 3 Model (Reinforcement Learning) extracted from [20]

2.4 Reinforcement Learning

An environment takes the agent's current state s_t at time t and action a_t as input, and returns the agent's reward $r(s_t, a_t)$ and next state s_{t+1} (see Fig. 3). The agent's goal is to maximize the expected cumulative reward over a sequence of action.

An agent interacts with an environment. Given the state(s_t) of the environment at time t, the agent takes an action a_t according to its policy $\pi(a_t|s_t)$ and receives a reward $r(s_t, a_t)$ (see Fig. 3). The objective of Tower of Hanoi is to find the solution in a way that is the shortest possible movement. To do this, we take actions that maximize the future discounted rewards. We can calculate the future rewards $R_t = \sum_{t'=t}^{T} \gamma^{t'-t} r_{t'}$, where γ is a discount factor for future rewards. In this article, the way of optimal solution is taken by the maximum action-value function:

$$Q(s_t, a_t) = Q(s_t, a_t) + \alpha \left(r_{t+1} + \gamma \max_a Q(s_{t+1}, a) - Q(s_t, a_t) \right) \qquad (6)$$

where $\alpha \in [0, 1)$ is the learning rate sequence, and γ is the discount factor.

3 Model

In this article, CNN and LSTM-RL methods are introduced. Our method is divided into two parts. The first method consists of extracting the features and predicting the multi-label by means of Convolutional Neural Networks method. The second method combines Long Short-Term Memory Recurrent Neural Network and Reinforcement Learning methods to obtain the shortest solution path.

Table 1 The configuration of Convolutional Neural Networks with 4 layers

Input image	$80 \times 80 \times 3$
Convolutional layer 1	$76 \times 76 \times 16$
Max-pooling layer 1	$38 \times 38 \times 16$
Convolutional layer 2	$34 \times 34 \times 32$
Max-pooling layer 2	$17 \times 17 \times 32$
Convolutional layer 3	$13 \times 13 \times 64$
Max-pooling layer 3	$6 \times 6 \times 64$
Convolutional layer 4	$2 \times 2 \times 128$
Max-pooling layer 4	$1 \times 1 \times 128$
Output	20×1 (Sigmoid)

3.1 CNN Multi-label Categorization

Table 1 shows our proposed architecture. This model with 4 layers takes a input image $80 \times 80 \times 3$ (3:color image,1:black & white image) and predicts a 20 dimensional tag vector.

The classifier was optimized by splitting 1186 original images into 9% training images (1067 images) and 10% validation images (119 images). We then tested the accuracy of the model with 300 test data. We labeled all the acquired images. For instance, given an image of participant (see Fig. 5), we would like this image to provide categories such as LD1, MD1, RD2, and TD1. A description of the multi-category is as follows: LD1-LD4 (There are one to four disks on the left side), MD1-MD4 (There are one to four disks in the middle), RD1-RD4 (There are one to four disks on the right), TD1-TD4 (Take disk 1, disk 2, disk 3, disc 4), PD1-PD4 (Put disk 1, disk 2, disk 3, disk 4). The per-class precision is shown in Fig. 4.

3.2 Combined Recurrent Neural Network and Reinforcement Learning

Based on the multi-label classification of CNN, we predicted possible action between sequential images (see Fig. 6).

In order to get the answer to the Tower of Hanoi, it is necessary to draw out possible actions in successive images. Table 2 shows the first six images of a participant and the probability of actions as possible. If there are four disks on the right side (RD4), then the probability of taking disk 1(TD1) is 45%. By the sixth image, the right disk shrinks to three and the probability of placing the disk 1(PD1) increases to 14 %. In the seventh image, the probability of placing the disk1 (PD1) in the middle increases

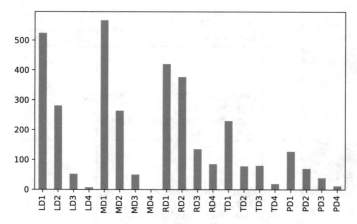

Fig. 4 The per-class barplots of input images. A description of the multi-category is as follows: LD1-LD4 (There are one to four disks on the left side), MD1-MD4 (There are one to four disks in the middle), RD1-RD4 (There are one to four disks on the right), TD1-TD4 (Take disk 1, disk 2, disk 3, disc 4), PD1-PD4 (Put disk 1, disk 2, disk 3, disk 4)

Fig. 5 There is one disk on the left (LD1) and the middle peg (MD1), and two disks on the right peg (RD2). The participant is holding d_1 (TD1)

to 27%. In this case, according to the coding rules of Fig. 7, we move the disk from the right to the middle, so we label it G4.

After predicting actions from sequential images, LSTM models can be optimized to predict observations and immediate rewards. On the other hand, RL models can be trained to maximize long-term rewards. We can calculate the probabilities distribution of the observation over all possible actions. These calculated probabilities are helpful for determining the next action. As shown in the Fig. 7, we can have two possible next actions {G1, G3} at time 0 according to the rules of TOH (see Table 3).

Table 2 Sequential images and probabilities

Sequential image	Multi-label and probability	Predictive action
	['**RD4**', '**TD1**', 'TD2', 'TD3', 'PD1', 'PD2', 'MD1', 'LD1'] ['**1.00**', '**0.45**', '0.01', '0.00', '0.00', '0.00', '0.00', '0.00']	TD1
	['**RD4**', '**TD1**', 'TD2', 'TD3', 'PD1', 'PD2', 'MD1', 'LD1'] ['**1.00**', '**0.45**', '0.01', '0.00', '0.00', '0.00', '0.00', '0.00']	TD1
	['**RD4**', '**TD1**', 'TD2', 'TD3', 'PD1', 'PD2', 'MD1', 'LD1'] ['**1.00**', '**0.44**', '0.01', '0.00', '0.00', '0.00', '0.00', '0.00']	TD1
	['**RD4**', '**TD1**', 'TD2', 'TD3', 'PD1', 'PD2', 'MD1', 'LD1'] ['**1.00**', '**0.44**', '0.01', '0.00', '0.00', '0.00', '0.00', '0.00']	TD1
	['**RD4**', '**TD1**', 'TD2', 'TD3', 'PD1', 'PD2', 'MD1', 'LD1'] ['**1.00**', '**0.44**', '0.01', '0.00', '0.00', '0.00', '0.00', '0.00']	TD1
	['**RD3**', '**TD1**', '**PD1**', 'TD2', 'LD1', 'MD1', 'PD2', 'TD3'] ['**0.98**', '**0.26**', '**0.14**', '0.11', '0.04', '0.04', '0.01', '0.01']	TD1
	['**RD3**', '**MD1**', '**PD1**', 'TD1', 'TD2', 'PD2', 'TD3', 'TD4'] ['**0.99**', '**0.92**', '**0.27**', '0.17', '0.15', '0.02', '0.01', '0.01']	PD1 **G4**

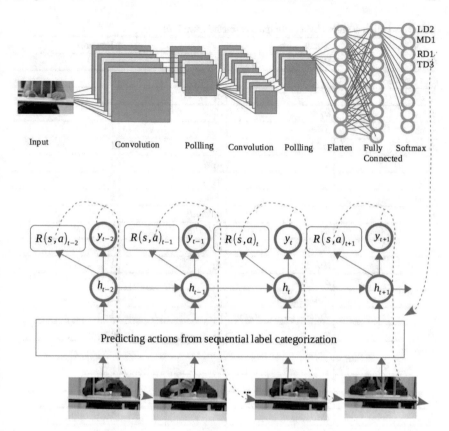

Fig. 6 Solving of Tower of Hanoi puzzle with CNN and LSTM-RL: x_t is the observation, h_t is the hidden state for LSTM, y_t is the predicted observation for time $t + 1$, $R(s, a)_t$ is the predicted reward

Table 3 Rewards and possible actions at time 0 extracted from [20]

$G1$	$G2$	$G3$	$G4$	$G5$	$G6$
1	0	1	0	0	0

Table 4 Probabilities and possible actions at time 0 extracted from [20]

$G1$	$G2$	$G3$	$G4$	$G5$	$G6$
0.35	0.003	0.60	0.00	0.001	0.001

Table 4 is achieved by looping an output of the network at time 0 with the input of the network. Therefore, base on the Tables 3 and 4, we choose the next action $G3$ (From Peg A to Peg C).

Table 5 Solution acquired by a participant extracted from [20]

	States		Rewards
Peg A	Peg B	Peg C	
$d_1/d_2/d_3/d_4$			0
$d_2/d_3/d_4$		d_1	1
d_3/d_4	d_2	d_1	1
d_3/d_4	d_1/d_2		1
d_4	d_1/d_2	d_3	1
d_4	d_2	d_1/d_3	1
d_1/d_4	d_2	d_3	-1
d_1/d_4		d_2/d_3	1
d_4		$d_1/d_2/d_3$	1
	d_4	$d_1/d_2/d_3$	1
d_1	d_4	d_2/d_3	1
d_1	d_2/d_4	d_3	1
	$d_1/d_2/d_4$	d_3	1
d_3	$d_1/d_2/d_4$		1
d_3	d_2/d_4	d_1	1
d_2/d_3	d_4	d_1	1
$d_1/d_2/d_3$	d_4		1
$d_1/d_2/d_3$		d_4	10
d_2/d_3		d_1/d_4	10
d_3	d_2	d_1/d_4	10
d_3	d_1/d_2	d_4	10
	d_1/d_2	d_3/d_4	15
d_1	d_2	d_3/d_4	15
d_1		$d_2/d_3/d_4$	18
		$d_1/d_2/d_3/d_4$	20

Table 6 Negative Reward and possible action extracted from [20]

G1	G2	G3	G4	G5	G6
-1	1	-1	0	0	0

Furthermore, Table 5 is a result acquired by a participant. This participant moved the same disk in a row (Between 6th and 7th line). In this case, the agent receives a negative rewards (−1) because this is not the optimal solution. Thus, in order to find the optimal solution, we modify selection algorithm by combining the calculated probabilities and rewards for possible action.

Table 7 Probabilities and possible actions extracted from [20]

$G1$	$G2$	$G3$	$G4$	$G5$	$G6$
0.0002	0.001	0.83	0.13	0.0004	0.002

Fig. 7 Coding—Move a disk from the left peg to the middle peg ($G1$). Move a disk from the middle peg to the right peg ($G2$). Move a disk from the left peg to the right peg ($G3$). Move a disk from the right peg to the middle peg ($G4$). Move a disk from the middle peg to the left peg ($G5$). Move a disk from the right peg to the left peg ($G6$) extracted from [20]

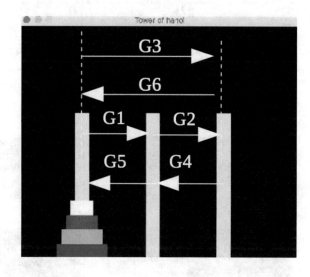

Based on the possible actions, we can predict possible action $G3$ in Table 7. But in order to maximize its cumulative reward (see Table 6), our LSTM+RL model takes an action $G2$ instead of $G3$.

4   Experiments

The participants were given four disks of the Tower of Hanoi that they had to solve. The instructions were given to the participants. The participants were requested to solve the four disks TOH as we collected their gesture. We took several videos of the processes by which the participants solved the puzzle. And these videos were transformed into a sequential images.

4.1   Experiment: Tower of Hanoi

We recruited 14 participants (Average age 41, Standard Deviation = 8.51). The blind group consisted of 6 women and 1 man (Average age 39, Standard Deviation = 6.65). The sighted group consisted of 6 women and 1 man (Average age 43, Standard Deviation = 10.30). Through these research experiments, we have obtained the

Table 8 Results of Tower of Hanoi for 15 participants extracted from [20]

Participant	Number of moves	Participant	Number of moves
1	15	9	44
2	25	10	35
3	21	11	23
4	48	12	38
5	23	13	15
6	24	14	34
7	32	15	26
8	30		

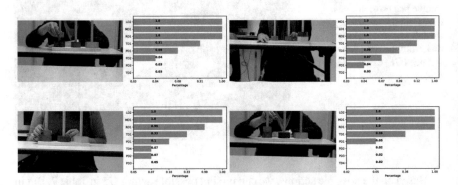

Fig. 8 The predicted result based on CNN

sequential data concerning about the solution of Tower of Hanoi. Table 8 shows results of Tower of Hanoi for all participants.

More specifically, the sighted participants made use of their deictic gesture which is used as grounding for a mapping between the object imagined and action. The deictic gesture forces them to remember what they have done in previous attempt. The result shows that the number of deictic gestures for this group is correlated with the total duration [$r = 0.44$, $p < 0.019$]. Meanwhile, the blind people build their mental representation with their hands trough touch. For the blind participants, the gestures added action information to their mental representation of the tasks by touching the disk or rotating it. the number of gesture for the blind people is correlated with the total duration. [$r = 0.496$, $p < 0.0072$]

4.2 Experiment: Multi-label Categorization

As shown in previous section, We have acquired a series of data from participants solving the Tower of Hanoi problem. These data contain semantic information,

Table 9 Results on performance metrics for all labels

Label	Precision	Recall	F1
One disk on the left(LD1)	0.9389	0.9435	0.9406
Two disks on the left(LD2)	0.9610	0.9716	0.9662
Three disks on the left(LD3)	0.8912	0.8289	0.8570
Four disks on the left(LD4)	0.4958	0.5000	0.4979
One disk in the center(MD1)	0.9242	0.9225	0.9233
Two disks in the center(MD2)	0.9588	0.9706	0.9645
Three disks in the center(MD3)	1.0000	1.0000	1.0000
Four disks in the center(MD4)	1.0000	1.0000	1.0000
One disk on the right(RD1)	0.8684	0.8618	0.8650
Two disks on the right(RD2)	0.9552	0.9490	0.9520
Three disks on the right(RD3)	0.9333	0.9906	0.9595
Four disks on the right(RD4)	0.9955	0.9444	0.9683
Take d_1(TD1)	0.5457	0.5535	0.5495
Take d_2(TD2)	0.5276	0.5454	0.5364
Take d_3(TD3)	0.5580	0.7201	0.5825
Take d_4(TD4)	0.4874	0.5000	0.4936
Put d_1(PD1)	0.5753	0.6312	0.5877
Put d_2(PD2)	0.4906	0.5127	0.5014
Put d_3(PD3)	0.4916	0.5000	0.4958
Put d_4(PD4)	0.4958	0.5000	0.4979

including actions and interactions with objects. In this experiment, given a series of data that are all labeled with a multiple categories, we have predicted these categories for a novel set of test data. For the evaluation metrics, precision/recall/F1 score have been used to evaluate the performance of a classification problems.

For precision/recall/F1 score, if the estimated label confidence for a label is greater than 0.5, the label is expected to be positive. However, for the estimated label confidence for the labels (TD1-4 and PD1-4), only positive label were assigned to the variable with the highest estimated label confidence of the eight labels. As Based

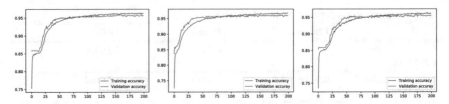

Fig. 9 Accuracy comparison for a convolutional neural network (CNN) with layers and epochs. The image on the left are 2 convolutional layers with 200 epochs. The middle is 3 convolutional layers with 200 epochs. The last image on the right is 4 convolutional layers with 200 epochs

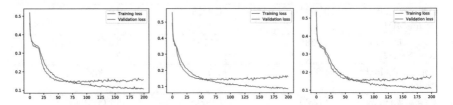

Fig. 10 Loss comparison for a convolutional neural network (CNN) with layers and epochs. The image on the left is 2 convolutional layers with 200 epochs. The middle image is 3 convolutional layers with 200 epochs. The last image on the right is 4 convolutional layers with 200 epochs

on the trained CNN model, we predicted the multi-label categorization for the test image. Figure 8 shows an example of the testing result.

According to the Table 9, the resulting F1 values for status labels(LD1-4,MD1-4,RD1-4) , which represented the model accuracy for multi-label prediction of CNN model, were high (>0.90) with the exception of LD3,LD4, and RD1 (0.8570,0.4979, and 0.8650, respectively). For the action's labels (TD1-4,PD1-4), a low degree of F1 value was found for TD4,PD3, and PD4 (0.4936, 0.4958, and 0.4979, respectively). We present visualizations of the learning curves (see Figs. 9 and 10). Overfitting occurred in the early 10 to 50 iterations. After 50 iterations, accuracy increasing with the number of iterations and training accuracy above the validation accuracy.

4.3 Experiment: LSTM+RL

The output vector obtained from CNNs contains semantic information. We feed this output vector into the RNN as an initial state for label prediction. And then, our model predict a possible action between sequential action. To do this, we conducted experiment by training our combined model(LSTM+RL) on the sequential data obtained in previous experiments on TOH solution. First of all, we evaluate and compare the performance of LSTM model on this sequential data. And then the combined model (LSTM+RL) is carried out.

The weights of all networks are initialized to random values uniformly distributed in the interval from $[-1/\sqrt{n}, 1/\sqrt{n}]$,where n is the number of incoming connections. To train our model we minimize the loss function for our training data.

To train the combined model we first initialized all of the LSTM's parameters with the uniform distribution between –0.1 and 0.1. We used stochastic gradient descent, with a fixed learning rate of .5. After 5 epochs, if loss increased in every epoch, we adjusted the learning rate. This LSTM model made predictions representing probabilities of the next action. To train the Q-function we initialized all Q-values of all state-action pairs to zero and initialized the states with their given rewards. Based on all possible actions obtained by LSTM, we measured a reward value for each possible action. If an action has the highest probability and reward, we can choose this as next sequential action. Otherwise we search another possible action with the highest reward value as next sequential action. Then, we updated the Q-value according to the equation (6) and repeated the process until a terminal state was reached.

4.4 Results

After training for the model RNN, we obtained the following shortest path : G1,G3,G2,   G1,G4,G5,G4,   G1,G3,G2,G5,G4,G3,G5,G1,G6,G5,G2,G1,G3,G2 (21 movements). The training error is shown in Fig. 11.

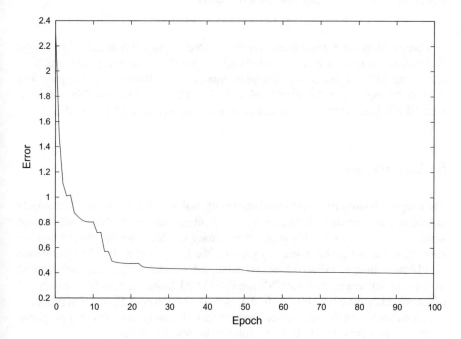

Fig. 11 LSTM train error

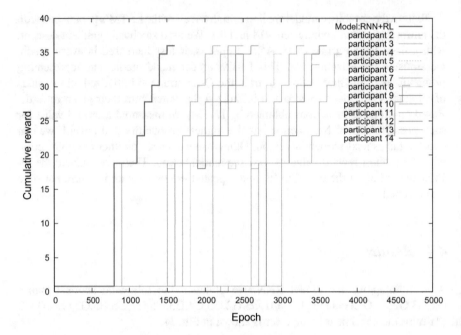

Fig. 12 Cumulative reward graph for LSTM+RL model

Compared to the experimental results in Table 8, this RNN model shows good performance improvement. Nevertheless, this result is not the fastest solution. According to the Table 8, the first participant and the thirteenth participant find the fastest solution. On the other hand, from Fig. 12, we can see that this combined model(LSTM+RL) improves the performance compared with LSTM model.

5   Conclusions

The images obtained from real world have many and complex categories. The objects around us, surrounding scenes, actions, and interactions with objects all contain semantic information. In this paper, we focused on the semantic information we encountered in the problem-solving process. We firstly explored multi-label image classification in order to analyze these semantic information. In addition, we proposed a convolutional neural network(CNN) and LSTM-RL based method that is capable of efficiently extracting features and finding solution of Tower of Hanoi. The proposed model has achieved the state-of-art performance. It can be seen that our proposed model can be a good choice to find a solution of Tower of Hanoi.

Further, there are two things that need to be studied in depth. First of all, as participants have a difficulty triggering simulations with visual object, deictic gesture for the sighted participants plays a central role in generating visual inferences. Meanwhile, the blind participants have a difficulty solving TOH because of the lack of tactile object. The interaction gestures for these participants play an important role in building their mental representation. Based on these observations, what we are wondering is how artificial intelligence constructs mental expressions like human beings and how to understand the meaning of such hand motions. The second thing to study is extremely simple. In the end, in order to find the solution of problem as like human beings, we must understand the rule of game accurately by understanding natural language. One disk at a time should be moved, in a location, a large disk cannot be placed on top of a smaller one. When someone plays the game directly by explaining the rules of the game as above, we can understand the rule of game while mapping languages and images. Likewise, we will do research to understand game rules through natural language processing and image analysis.

References

1. Krizhevsky, A., Sutskever, I., Hinton, G. E.:Imagenet classification with deep convolutional neural networks. In: NIPS (2012)
2. Barsalou, L. W.: Grounded Cognition: Past, Present, and Future. Topics in Cognitive Science, 2(4), 716–724 (2010). http://doi.org/10.1111/j.1756-8765.2010.01115.x
3. Chan, T.-H.: A statistical analysis of the towers of hanoi problem. International Journal of Computer Mathematics, 28(1-4), 57–65 (2007). http://doi.org/10.1080/00207168908803728
4. Cho, K., van Merrienboer, B., Gulcehre, C., Bahdanau, D., Bougares, F., Schwenk, H., Bengio, Y.:Learning Phrase Representations using RNN Encoder–Decoder for Statistical Machine Translation (pp. 1724–1734). Presented at the Proceedings of the 2014 Conference on Empirical Methods in Natural Language Processing (EMNLP), Stroudsburg, PA, USA: Association for Computational Linguistics (2014). http://doi.org/10.3115/v1/D14-1179
5. Eck, D., Di Studi Sull Intelligenza, J. S. I. D. M., 2002. (n.d.): A first look at music composition using lstm recurrent neural networks. People.Idsia.Ch (2002)
6. Elman, J.L.: Finding Structure in Time. Cognitive Science **14**(2), 179–211 (1990). https://doi.org/10.1207/s15516709cog1402
7. Nathan, M. J.: An embodied cognition perspective on symbols, gesture, and grounding instruction. Symbols (2008)
8. RICHARD, J.-F., Poitrenaud, S., Tijus, C.: Problem-Solving Restructuration: Elimination of Implicit Constraints. Cognitive Science, 17(4), 497–529 (1993)
9. Sutskever, I., Hinton, G.E., Taylor, G.W.: The Recurrent Temporal Restricted Boltzmann Machine **1601–1608**, (2009)
10. Sutskever, I., Martens, J., Conference, G. H. P. O. T. 2. I., 2011. (n.d.). Generating text with recurrent neural networks (2011)
11. Varela, F., Thompson, E. (n.d.). ,E. Rosch. 1991. The embodied mind: Cognitive science and human experience. Cambridge (1991)
12. Szegedy, C., Liu, W., Jia, Y., Sermanet, P., Reed, S., Anguelov, D., et al. (2015). Going Deeper With Convolutions. Cv-Foundation.org , 1–9
13. Russakovsky, O., Deng, J., Su, H., Krause, J., Satheesh, S., Ma, S., et al.: ImageNet Large Scale Visual Recognition Challenge. International Journal of Computer Vision **115**(3), 211–252 (2015). https://doi.org/10.1007/s11263-015-0816-y
14. LeCun, Y., Bottou, L., Bengio, Y.: Gradient-Based Learning Applied to Document Recognition. Proceedings of the IEEE **2278–2324**, (1998)

15. Gong, Y., Jia, Y., Leung, T., Toshev, A., and Ioffe, S. (2013, December 17). Deep Convolutional Ranking for Multilabel Image Annotation. arXiv.org
16. Oquab, M., Bottou, L., Laptev, I., and Sivic, J. (2014). Learning and Transferring Mid-Level Image Representations using Convolutional Neural Networks, 1717–1724
17. Sharif Razavian, A., Azizpour, H., Sullivan, J., and Carlsson, S. (2014). CNN Features Off-the-Shelf: An Astounding Baseline for Recognition, 806–813
18. Girshick, R., Donahue, J., Darrell, T., and Malik, J. (2014). Rich Feature Hierarchies for Accurate Object Detection and Semantic Segmentation, 580–587
19. Sermanet, P., Eigen, D., Zhang, X., Mathieu, M., Fergus, R., LeCun, Y.: December 21). Integrated Recognition, Localization and Detection using Convolutional Networks, OverFeat (2013)
20. Bang, S., and Tijus, C. (2018). Problem Solving using Recurrent Neural Network based on the Effects of Gestures. In: 10th International Joint Conference on Computational Intelligence (IICCI), pp.211–216. https://doi.org/10.5220/0006931802110216
21. Hochreiter, S., and Schmidhuber, J. (2006). Long Short-Term Memory. Dx.Doi.org, 9(8), 1735–1780. https://doi.org/10.1162/neco.1997.9.8.1735

Author Index

© The Editor(s) (if applicable) and The Author(s), under exclusive license
to Springer Nature Switzerland AG 2021
C. Sabourin et al. (eds.), *Computational Intelligence*, Studies in Computational
Intelligence 893, https://doi.org/10.1007/978-3-030-64731-5

Printed in the United States
by Baker & Taylor Publisher Services

Printed in the United States
by Baker & Taylor Publisher Services